AF837521

The Value of Water in a DRYING CLIMATE

Editors: Tor Hundloe, Bond University and Christine Crawford, University of Tasmania

PUBLISHING

© Tor John Hundloe and Christine Crawford 2012

All rights reserved. Except under the conditions described in the *Australian Copyright Act 1968* and subsequent amendments, no part of this publication may be reproduced, stored in a retrieval system or transmitted in any form or by any means, electronic, mechanical, photocopying, recording, duplicating or otherwise, without the prior permission of the copyright owner. Contact **CSIRO** PUBLISHING for all permission requests.

National Library of Australia Cataloguing-in-Publication entry

The value of water in a drying climate/edited by Tor Hundloe and Christine Crawford.

9780643101609 (pbk.)
9780643106635 (epdf)
9780643106642 (epub)

Water-supply – Australia.
Water demand management – Australia.
Climatic changes – Australia.

Hundloe, T. J. (Torstein John)
Crawford, Christine.

333.91140994

Published by
CSIRO PUBLISHING
36 Gardiner Road, Clayton VIC 3168
Private Bag 10, Clayton South VIC 3169
Australia

Telephone: [+613] 9545 8555
Local call: 1300 788 000 (Australia only)
Fax: +61 3 9662 7555
Email: csiropublishing@csiro.au
Web site: www.publishing.csiro.au

Front cover: Little Swanport estuary (photo by Francisco Neira)

Set in 10.5/12 Adobe Minion Pro and Stone Sans
Edited by Elaine Cochrane
Cover design by Alicia Freile, Tango Media
Text design by James Kelly
Typeset by Desktop Concepts Pty Ltd, Melbourne
Index by Indexicana
Printed by Ingram Lightning Source

The publication of this book was made possible through the generous support of the Tasmanian Aquaculture and Fisheries Institute (TAFI (now IMAS)) and the Fisheries Research and Development Corporation (FRDC).

CSIRO PUBLISHING publishes and distributes scientific, technical and health science books and journals from Australia to a worldwide audience and conducts these activities autonomously from the research activities of the Commonwealth Scientific and Industrial Research Organisation (CSIRO). The views expressed in this publication are those of the author(s) and do not necessarily represent those of, and should not be attributed to, the publisher or CSIRO.

Feb26_RP_ILS

Contents

List of contributors

Christine Crawford is a senior research scientist at the Institute of Marine and Antarctic Studies (formerly the Tasmanian Aquaculture and Fisheries Institute) at the University of Tasmania.

Peter Daniels is a Senior Lecturer in the Griffith School of Environment, Griffith University.

Barry Gallagher is a PhD student at the Institute of Marine and Antarctic Studies at the University of Tasmania.

John Gibson is an aquatic chemist with a long-held interest in studying lakes and estuaries, in particular at high latitude.

Tor Hundloe is Professor of Environmental Management and Science, Bond University; Adjunct Professor, Environment School, Griffith University; and Emeritus Professor, Environmental Management, the University of Queensland.

Jeff Ross is a research fellow at the Institute of Marine and Antarctic Studies at the University of Tasmania.

Michelle Wenner is a Policy Officer, Fisheries, Department of Primary Industries, Victoria.

Amy White is a Senior Teaching Fellow in the Institute of Sustainable Development and Architecture, Bond University.

Acknowledgements

We wish to whole-heartedly thank the people of the Little Swanport catchment, especially the Little Swanport Catchment Management Implementation Committee and the catchment members who completed our surveys, for their continued hospitality, interest, patience and support for the project.

We are also grateful to the Fisheries Research and Development Corporation (FRDC) and the Tasmanian Aquaculture and Fisheries Institute (now IMAS – Institute of Marine and Antarctic Studies) for their funding and support for the research and publication of this book.

A big thank you also to the people who supported us with our research, including Oyster Bay Oysters for their logistical support, to Sam Foster our ever reliable technical officer, to Mel Kelly and Sandy Dunbabbin in the catchment, to Dr Beth Fulton at CSIRO for advice on model development and to Ted Lefroy who chaired the initial steering committee and suggested the title of this book.

We have found our publisher, CSIRO Publishing, to be a source of inspiration, sound advice and much help. In particular we wish to thank Ted Hamilton who saw the value in our project. Elaine Cochrane undertook the critical task of editing the book and, in doing so, improved its presentation. Finally, to see the project through to completion we had the expert assistance of Tracey Millen. Their efforts are much appreciated. Any omissions or errors remain ours.

Prologue

The terrestrial world is a myriad of catchments. Only the person who stands on a mountain peak is temporarily not in a catchment. Once rain or snow falls, this person becomes in and of a catchment, regardless of which way the water runs.

Catchments provide the best possible ecological boundaries for the study of nature and of human–nature relationships. We have come to learn the hard way that political boundaries (both between nations and within nations) are a tremendous hindrance to the successful management of human interactions with the environment.

Air- and water-borne pollutants have no respect for the lines we humans draw on maps. As a consequence we have, for decades now, experienced the fall of acid rain in Scandinavia from the chimneys in distant parts of industrialised Europe. There are polluted river mouths in many developing countries where rivers gather human, agricultural and industrial effluent on the way to the sea. The rivers carrying the pollution load are likely to have run through a number of countries, each upstream country hoping to have its pollution out of sight and therefore out of mind. Trans-boundary pollution has been common for decades.

A new threat exists today. Only recently have we come to recognise the probable effects of climate change. It is very early days in terms of our understanding of climate change and the future is uncertain. In the case of global pollutants such as the greenhouse gases, catchments are not the appropriate boundary; the planet is clearly one ecosystem when the pollution is global.

The great oceans are global commons as much as the atmosphere is. Trace some of the great rivers through the industrialised parts of the globe and witness their increasing deterioration as the factories, farms and households, country after country, deposit their wastes, before the river with its load of rubbish enters an ocean. It is not only pollution that is a trans-boundary problem for water. There is the matter of allocation. The fresh, drinkable water on which all life depends is fought over, with the upstream countries better placed to take a far greater share than those downstream. We are without protocols to manage these matters.

It is one thing to recognise that catchments form natural boundaries; it is another to understand, measure, and manage human activities on a catchment basis. For about 150 years (the age of most modern nations) we have attempted to manage human activities on a national basis – with significantly varying degrees of success. We keep attempting to improve the management of our affairs, including human–environmental relationships, within our borders. Even if we perfected management at this level, we would leave the trans-boundary environmental problems unresolved.

In this book we are journeying into the uncharted waters (no pun intended) of describing, understanding and modelling human–environment relationships in a catchment. What we set out to do is not some theoretical exercise; rather it is what is required if we are to make best, sustainable use of nature's gifts. How we manage human affairs within ecosystem boundaries is our challenge.

This book gradually evolves to become the story of the everyday life of one small Australian catchment. We will attempt to treat the catchment as if it were an autonomous unit, a pretend nation, with its inhabitants needing to ensure they make optimal use of their environment – which means making best use of the available water. The later chapters that take on this role and the ones that lead up to it, and set the scene for it, are about people in and of the environment. These are people earning an income from the land, soil and water that comprise the key natural resources of river basins. It is about people living in rural settings in upper catchments with soils and rainfall suitable for farming; people residing in coastal settings in the lower catchments suitable for residential 'lifestyle' villages; people in towns and cities; and people working and relaxing in estuaries where fishing and aquaculture are likely to occur.

The idea of a catchment is universal. At long last – approaching 50 years since Rachel Carson wrote *Silent Spring* – we are ready to treat human–environment interrelationships as fundamental to our well-being and future. We have been struggling to find sensible physical and social boundaries to target our governance. One lake is too small; a large island (such as Australia) or a continent (again think of Australia) is too big. A catchment – or river basin – is the sensible unit waiting to be recognised.

No two catchments are the same. Hence this book is not simply a biography of one catchment, although later on we do focus on a small Australian catchment in the island state of Tasmania. It is possible to make a range of important statements about life in a catchment – any catchment. First and foremost we tend to think of catchments (or river basins, drainage basins or watersheds as they are alternatively called) because they actually 'catch' rain and snow from the atmosphere and eventually end up with water running through streams to an outlet into oceans or, in some cases, internal lakes. The area that catches the rain or snow defines the boundary of the catchment. Humans and the other animals and plants with which we share the planet maintain their lives by accessing the water that enters and flows through catchments. Hence our study will be the story of water as much as the life it sustains.

Our story will take us back in time. This is for number of good reasons. First, we will seek to understand how our ancestors thought of water. Never dismiss our ancestors' beliefs as irrelevant. The Hindu, Buddhist, Confucian and Jewish belief systems remain as strong today as they did in ancient times, notwithstanding all the fantastic inventions – based on science – that we take for granted. One thing we learn from the social sciences is the enormous disconnect that can exist between faith and science, and the problems this causes. When faith (superstition if you like) rules, the scope for a sound scientific decision is very limited, often impossible.

From our brief history of human attitudes to and interest in the use of water we will move to describe some of our world's largest catchments, again with attention on the human dimension. We do this because we seek to learn.

The critical questions throughout this book will be: are we making best use of our water, the land it feeds and the natural resources dependent on both the water and land? How should we judge this? The introductory chapter sets us on that road. It is a journey that inevitably takes us to the twenty-first century challenge to achieve sustainable development – how to live sustainably. To make sense of what this entails we outline the principles of sustainability in Chapter 2. The fundamental principle of sustainability is the marriage of economics and ecology, and that leads to a discussion in Chapter 3 of how economists attempt to value nature. We will discover the difficulty, if not impossibility, of deriving values we can justify on ethical grounds. Only when we have these background chapters behind us can we commence our discussion of water and the world's great catchments.

Our case study is based on a small Tasmanian catchment. Its small size is very important. We have been able to get to know virtually every tributary of the river around which the catchment is based – the river that has formed the catchment. The fact that only a few hundred people live in the catchment means that we have come to meet, and talk to, a significant number of the people in the catchment and we have come to understand how they make a living from the resources of the catchment. From this we obtain a measure of the monetary value of the natural and human resources in the catchment. This is where we analyse our catchment as we would a small nation-state, exporting, importing and self-sufficiency. If only we had the data to undertake this type of exercise for the world's largest catchments! Imagine our knowledge and, based on that, our foundation to make better human–nature decisions.

We are interested in more than the equivalent to the catchment's gross domestic product (what economists would term the regional product). In addition to this measure, we are seeking to discover and document values that are not recorded in the market. We need to expand economics, and economic thinking, if our goal of sustainability is to be given a chance of success. We will attempt to measure the values that to date have been neglected.

In addition to and separate from these measures, we have been able to delve into the values people put on their lifestyle. Every lifestyle is unique, and we recognise that it will be difficult to generalise from the individual to society's values. However, much in our study is of universal relevance. The story we tell and the analyses we undertake are ones that will help anyone interested in human–environmental interactions understand what they need to comprehend to ensure healthy catchments and healthy people. This is our aim.

The book is a guide: first and foremost about how we should think about ecology, economics and ethics – using the extremely sensible ecological boundary provided by a catchment to arrange our thoughts and direct our analyses. It is, to our knowledge, the first attempt to use a catchment to illustrate how we can better understand human–environment relationships using the twenty-first century paradigm of sustainability science.

Tor Hundloe

1

Introduction

Tor Hundloe

Ask people to think of a catchment and they tend to think of large geographical areas and extensive communities, for example the Nile River and all the people who work on and around it and benefit from it. As the Nile catchment illustrates, many large catchments extend beyond the boundaries of one nation. For some of the world's largest river basins, not just two, but many more countries can have some or all of their territory in the basin.

National boundaries dissect catchments. A variety of different land uses and human occupations make diversity the norm of catchment economies and society. Notwithstanding, people's everyday reliance on the world's great rivers and their catchments override all else. The water in the basin links all. Without it there is none of what makes a catchment a vibrant ecological and human entity. This reality matters much more than the political boundaries that divide so many catchments.

Where humans commenced the journey from hunter-gatherer to farmer, and finally to urbanised citizen, catchments came into their own. Hunter-gatherers had to range far and wide to find food; hence they did not concentrate in large numbers in any one place and they did not put down roots. Once farming was discovered, decisions had to be made about the best place to grow plants and to corral domesticated animals. River catchments provided the natural environments for this radical lifestyle change. As the surplus of farming became available for the expansion of human activities, towns and then cities formed. For obvious reasons, the lower reaches of major rivers provided the ideal location for the formation of cities, and with cities came civilisations.

Civilisation is what we have come to call the economic, cultural and social organisation of people into complex societies where surpluses of food and other goods are generated for communal uses, and the division of labour underpins work. The process from discovering the advantages of growing crops on flood plains, or at least near to flowing water, to eventually centring huge cities on river banks, if not near agriculture on river plains and river deltas, was not necessarily rapid and certainly not smooth. Success came and went. To underscore this, some of the world's once very productive agricultural catchments are now near-deserts. Humans managed to destroy productive soils through over-use. They managed to starve flood plains of water by upstream diversions of water to those with the political power to command it. They attempted to irrigate far too much land with limited water flows. These were the mistakes of the ancient world, the results which we can see today in the deserts of once-productive agricultural and urban civilisations in many places in the Middle East.

Not all our ancestors were fortunate enough to live in natural environments that provided them with the resources they could marshal if they put their minds to it. It was the challenges of how to benefit from nature's free gifts that played on the human mind and led to invention becoming a defining human characteristic. Farming itself was such an invention. The challenge was to provide food season-in, season-out, year-in, year-out, and overcome the vagaries of an uncaring climate – god or nature. The idea of saving seeds and replanting them was a small step in solving this challenge. Not all were in a position to undertake such daring experiments. They were daring because if the expected crop did not grow, people who had become set on being farmers would be forced to revert to wandering (in a relative wilderness at least) as hunter-gatherers in search of the next meal.

It was beyond human ability to convert the recalcitrant soils of arid environments to agriculture; hence, the ever-wandering peoples of the north African and Arabian deserts and the hunter-gatherer existence of the Inuit, the Australian Aborigines and other geographically unlucky people – unlucky because there was nothing in their environment to suggest the possibility of agriculture in the form in which it developed on the river banks of the great Chinese, Indian and Middle Eastern rivers. In these richly endowed areas, water-sufficient environments determined economic and social evolution. In these places humans had the right plants and animals, soils and rainfall to work with. The outcome is the society of the twenty-first century. The world and its people are what they are today due to our forebears being able to develop the sciences of agriculture, engineering, accounting and political organisation, something which has eluded all other animals. With this success comes an enormous responsibility for the human animal.

Until the Industrial Revolution commenced about 250 years ago, geographical determinism was reality. Before we had the tools of modern manufacture, we relied on our natural inheritance, the soils, the rainfall and the animals we could domesticate to assist us farm. All of us might have come out of Africa, but where we eventually settled (fertile river plain, desert, or tundra) made all the difference to how we learned to live and how we continue to live. This makes it appear too clearcut. It wasn't. The interested reader will find Jared Diamond's *Guns, Germs and Steel* one of the most enlightening books on human–environmental relationships as they formed our early history.

In the very early farming communities, the ability to put aside a surplus led to agriculture and invention, and the next invention built on the ones before. A process of material progress was underway. The vast array of the basic tools used in farming were each and every one of them an invention. The time and resources involved in inventing things, of whatever kind, required a surplus of food, enough that a few would be relieved from the daily grind of farming to tinker away as 'inventors'. This social relationship necessitated investment; investment being the means to provide food, shelter and other material requirements from the produce of past work. People no longer lived day-to-day, but thought of the future – a better future – and planned for it by investing, the most basic means being the putting aside of a few seeds to plant next season, and from the harvested crop putting aside some food for the non-farmer, the artisan who would (with luck or much endeavour) invent the next tool for cultivating the soil.

Farming was undoubtedly the earliest form of investment – working hard, putting aside seeds for the future, and then finding oneself with a new and potentially larger crop next season. The productive soils, particularly in the flood plains of lower catchments allowed for the provision of surplus foods and other resources that underpinned the human development we now take for granted in much of the world. Great civilisations grew from nature's original gifts of fertile river basins. We will describe these later. Those living in less productive environments were people who were at a disadvantage from the beginning and were destined to hunter-gatherer existence, some continuing to the present. Nature knows no notion of distributive ethics.

Not all early farming was to prove sustainable. In the early days of agriculture our knowledge of nutrients, chemicals and predator–prey relationships was seriously lacking. We stripped soils of nutrients, we killed large numbers of animals that had been effective biological controllers of the 'pests' that ate our farmed foods, we over-irrigated, and did much more of a negative nature. Until the invention of science by the ancient Greeks, which was only 2500 years ago, we based much of what we did on myth and superstition and hoped, even expected, the gods to deliver us water and good harvests. We gave our gods names that suggested their powers, to assist us with rain, with fertility, with good harvests. We blamed ourselves when they did not deliver, such was our primitive psychology. We must have offended the gods. Today we have considerable scientific knowledge, yet it seems that it is easier to pray for rain than order our lives around the knowledge we have accrued of the hydrological cycle.

Imagine a 'Garden of Eden'.[1] I think of a jungle in which there are snakes and leaves large enough to cover crucial human body parts, the leaves being from rain-gathering trees. Need there be an apple tree? Obviously this author has seen too many drawings of the mythical Garden of Eden. Such a place was supposed to exist some thousands of years ago, and some of us reckon we know where it was located. Turn on the television and see the Garden of Eden as an Iraqi desert that is being fought over in the later part of the first decade of the twenty-first century[2] – once the cradle of civilisation, now a possible foretaste of the end of it.

River basin history

Early human history is largely centred on great rivers. By great we mean long rivers, usually carrying plenty of water – some every day, and the others during floods. Think of the Nile, the Amazon, the Huang He (Yellow), the Yangtze, the Mekong, the Ganges, the Indus, the Volga, the Euphrates, the Tigris, the Mississippi, and the Murray–Darling. We will discover in a later chapter that the ones mentioned here are not all in the top 10 largest rivers; however, they are ones most readers would be familiar with. Some of the large and important rivers in Russia and Africa escape everyday attention. The Congo is one of these, as are those in central Russia. We will discuss these later.

The first known civilisations occurred on the flood basins of the Euphrates and Tigris rivers (the once fertile part of present-day Iraq). Following this, civilisations were formed on the Nile, the Huang He, the Indus and the Ganges. For example, there was a substantial Egyptian civilisation in the Nile Delta south of Luxor well before the pyramids were built – and that was a very long time ago.

Move forward to a far more recent era, and we discover that long rivers became the freeways of explorers, raiders and traders. The lengthy rivers of Europe were ideal passages for Vikings in their shallow-draught boats. The Viking era commenced in the tenth century. The Swedish Vikings travelled the length of the Volga, establishing the first Russian state (called Rus) in the process. These Viking ships were the equivalent of today's super aircraft carrier – state of the art technology. The Norse longships could wind their way for extensive distances in the shallowest of rivers just as easily as they could conquer the high seas, as Leif Ericsson proved when he crossed the Atlantic Ocean to America about a thousand years ago.

Move forward once again to a more recent period, to the 'wild west' days of North America, from the eighteenth century on to the early twentieth century. The web of large, long rivers in what was to become the United States provided a major means of transport for both products and people. The glamorous paddle-steamers are as much a part of North American history as the wheeled wagon, coach and cowboys. The wheelers and dealers, cardsharps and ladies who worked the boats were the waterways' equivalent of cowboys, outlaws and bar-room entertainers. The river folk were just as important in 'opening up' the United States as the cowboys and the wagon trains.

At about the same time as rivers were allowing the United States to be colonised by the newly arrived pastoralists and farmers, the British were digging canals to serve as transport corridors between the industrial midland cities and the coastal ports. Raw cotton came into Manchester and Liverpool from the colonies, and once on British shores moved inland to the mill cities via the canals. Manufactured garments found their way back to the colonies (and other markets) via canal-boats to British ports, then ocean-going ships. Water transport was the easiest, quickest and cheapest means of transport in the early days of the Industrial Revolution.

The case today

Come to the present. By the end of the twentieth century, most (if not all) of the great rivers have been tamed by massive dams for the purposes of irrigation and/or hydro-power. More often than not, particularly in the past 20 or so years, the construction of dams has been controversial and politically charged. Water allocations are altered when dams are constructed. Downstream uses can suffer as a result of seriously interrupted flows, and natural ecosystems can be damaged with the reduction in environmental flows. In poor countries, peasants have protested at the inundation of their land. In rich countries, 'not-in-my-backyard' (NIMBY) groups have often successfully stopped the damming of many a local river. These groups are quick to play the environment card, but self-interest is the fundamental motivation in many cases.

The benefit-cost analysis of altering water flows is generally a very complex calculation. There is ample evidence that many of the analyses undertaken to date have been flawed, and as a consequence water storages have been built when they should not have been, or have not been built where they should have been. We will say much more about water, the environment and economics in future chapters.

Nothing has been more controversial than when an upstream country constructs a dam, and as a consequence alters the flow into downstream countries. Wars have been the result. For example, an often overlooked aspect of the Palestine–Israel conflict is a dispute about access to water. The River Jordan rises in the Golan Heights in southern Syria, occupied by Israel since 1967, and is a significant part of the border between Israel and Jordan and the Palestinian territory of the West Bank and Jordan. Religion may be the face of armed conflict of this festering sore of Middle Eastern politics; however, control of resources is a major cause of the bloody conflict.

More water wars are on the horizon, particularly where rivers meander through a number of countries and where populations continue to increase and the need for basic economic growth is paramount. Rivers find their way through many countries in Europe and the Europeans have been able to settle on acceptable protocols. This is not necessarily the case in parts of Asia and Africa. It is not just quantities of water that lead to disputes; water quality is a major issue in some of these existing, or potential, conflicts.

Water quality was one of the main catalysts for the first United Nations conference on the environment, held in Stockholm in 1972. The dirty rivers flowing through country after country in industrialised Europe provided the evidence that trans-boundary pollution was the environmental problem of the era, and that international agreement was required to address it. In following years, the richer countries did much to clean up their rivers, to the benefit of all who relied on the particular rivers. Rich country governments can do this.

Other than that, and cleaning up trans-boundary air pollution in the richer parts of the world, little was to result from the Stockholm meeting of world leaders. We had to wait until 1987 and the report by the World Commission on Environment and Development to see water quality, water quantity and its allocation on a global scale put on the agenda, along with a range of other unresolved environmental issues. This was when the concept of sustainable development (or sustainability) took hold, first in government departments throughout the global community, then in the business world. Today the idea of sustainability is the Holy Grail. One hopes that those who place their hands on their hearts to swear by it understand it.

A paradigm shift: a new view of river catchments

By treating a water catchment as a geographic whole – regardless of the political and administrative units into which it is divided –our eyes are opened to a much more rational way of viewing the world than the artificial (political) perspectives we have become used to. Just for a moment, think of the countries that come to mind because a part (if not all) of their boundary with a neighbour is a river. The Mekong River, being a boundary between Thailand and Laos, is a good example. Both sides of the river depend upon its water flow for their agriculture and domestic water. The fishers who sit in their boats in the middle of the river could belong to either country.

The politics and the economics of a catchment are likely to differ significantly from those of a nation, and the types of human relationships also differ. Catchment residents, when they are forced to confront and live within nature's boundaries rather than within political ones, have a different perspective of their world. We will illustrate this with empirical data as we progress through the book.

Sustainability science

This book is not simply the story of river basins, although that in itself has potential to be a minor thriller. Too much blood and guts have been spilt over access to water for it not to be. However, our focus is not the exercise of political, economic – and military – power. The book is designed by scientists (a combination of natural scientists and social scientists) and hence focuses on scientific matters. We seek to solve intricate problems – ones that need to be solved before they escalate to conflicts. While researched and written by scientists, the book is designed for anyone who is interested in environment–human relationships. It opens up to the reader a new science, one that is going to dominate the twenty-first century.

The scholars we know of as environmental scientists have, ever so gradually, developed a radically new way to tackle their subject matter. Until very recently (40 years ago) we divorced ourselves from environment–human relationships. We studied botany, biology, hydrology and, more recently, ecology, believing that if we knew enough about these subjects we could protect nature and make a better world. We were wrong. While we certainly need to understand these sciences, we have come to realise that we must also come to know ourselves – our desires, our lifestyle choices, and our social and economic interrelationships. We have discovered that knowledge in, and application of, the social sciences (economics, psychology, sociology and others) is fundamental in meeting our environmental responsibilities. We attempt to bridge the gap between the natural and the social in this book. This places this book into a different category from the plethora of books on the environment.

The paradigm shift we propose, and illustrate, results in a new science, sustainability science. It has been fashioned by a deliberate marriage of ecology, economics and sociology, and within a specific ethical framework. This new science had its genesis with the release in 1987 of the report by the United Nation's World Commission on Environment and Development titled *Our Common Future*. The report, published in book form by Oxford University Press, is alternatively known as the *Brundtland Report* after the chair of the commission, Gro Harlem Brundtland, then Prime Minister of Norway. We will come to understand this new science better in the following chapters. To help in the comprehension of sustainability science we develop the concept in the next chapter.

2

Introduction to sustainability

Tor Hundloe

What do we think when using the term sustainability? If we wish to sustain the ecology, the economy, and the social relationships in a river basin, what do we mean? I imagine we think of it in terms of sustaining our particular lifestyles, assuming they are enjoyable. We think of sustaining jobs – no one wishes to worry about losing paid work. This would apply regardless of how we view our work – even work we do not enjoy is valuable because it pays the bills. We want to sustain our incomes however they are earned, as profits, as wages, as rents. And, of course, we need to sustain the ecosystems on which we rely for both income and pleasure, and more generally for life support. Without the gifts of nature we would not exist. This we have come to realise over the past 40 or so years.

You are entitled to think that all these propositions are common sense and that the concept of sustainability is not a new idea. You would be partly correct, but not totally. We have given too little thought to the relationship between the health of our ecosystems and the wealth of our economy and ourselves. We have tended to be short-sighted and narrowly focused, and hence we have produced considerable pollution, environmental degradation and loss of species, and are now threatening to cause serious climate change and extreme weather events (hurricanes, bushfires, snow storms, floods) through the build-up of atmospheric pollution (carbon dioxide and other greenhouse gases). We worry how our agricultural system can support an extra two to three billion people, which will be the need in 30 to 40 years. How we can ensure that there is enough fresh water where it is most needed with this increased global population?

Taken together, these assaults (if that is not a too strong word) have led to a very serious overshoot of the environment's carrying capacity if the analyses by the ecological footprinting experts are correct. They suggest we have already overshot by approximately 40 per cent. The overshoot is not deliberate – humans do not deliberately harm themselves – but rather a consequence of ignorance, short-sightedness and faulty economics, which puts pollution out of mind. There are only minimal caps on our pollution and we need more than a cap. The necessary claw back of greenhouse gas emissions is going to be very difficult to achieve politically. We could fail, and we are likely to fail if we are overly pessimistic. We must not contemplate failure. Because we are all at fault, we all have responsibilities from this day on.

We have been led by the wonders of technological advances to become far too optimistic with regard to finding new mineral and energy resources. The idea of a 'peak', as in 'peak oil', is not yet established in the public mind. The rapidly growing world population has become an unfashionable subject to discuss. Even those who are keen to call a halt have no practical plans for the countries where they are most needed. Population growth has stabilised and is even falling in mature, industrialised countries. On the other hand, all the aid programs have failed to reduce population growth in the majority of poor countries. The ever-increasing middle class in countries like China puts ever-increasing pressure on the planet. Yet, we must salute the growth of the middle class because global justice requires that all should be able to live the lifestyle we enjoy. If we can't all enjoy a healthy and relatively wealthy lifestyle, if only a certain number can live at that level, we must accept the Malthusian solution and work to reduce the global population over the long term – until we reach the desired level.

Our common future

We have recognised the seriousness of most environmental problems for some time now. As a consequence, there has been a shift in values, attitudes and practice, particularly in the richer countries and the middle class of the poorer ones. On certain environmental matters we are well on the way to the necessary understanding to achieve our goals. Localised pollution is easily recognised. Eco-toxins, once identified, can be banned without political ramifications.

We are coming to recognise that in the long term a healthy, sustainable economy depends on a healthy environment where degradation is not occurring, where natural resources are utilised on a sustainable basis, and where nonrenewable resources are replaced by investing the rents and profits in alternative energy sources (known in economics as the Hartwick principle). This is the very practical proposition that unless there is investment to compensate for the income-earning loss of nonrenewable fuels, minerals, and any other nonrenewable resources, future generations will be worse off. While we recognise this, we don't necessarily practise it, and consequently we will be worse off in the future in economic terms.

Theory and practice diverge. Because of this we need practical examples that illustrate the benefits of good theory. A major reason, certainly the most important one, for the writing of this book is to show *how* we can and should think about sustainability. We recognise that it is unlikely to provide all the answers. This for two reasons: one, our knowledge is limited; two, our disciplines are not easily meshed into a coherent sustainability framework. We will come to understand both these issues and contemplate their resolution as we progress, and that we hope will be of real benefit to those who follow in our footsteps.

It is not only the limits imposed by ecology that are fundamental to a way forward to sustainability. We must recognise that our economic actions – how we organise our production, how we regard and reward workers, resource owners, and investors – can have a significant impact on the type of society we form and in how we relate to the natural world. In the context of sustainability, our economy is a sub-unit of society, not the other way round. In other words, we cannot let the economy determine the type of society we live in. We will need to define the criteria for a sustainable economy.

There are the social characteristics of sustainability. What are they? On this issue we have a less comprehensive theory than we have for our ecological and economic relationship. While we can identify an overriding economic goal for society (on which we will say more later), we do not have one by which to define an overriding sociological goal. Is there an overriding goal or are there numerous competing – or conflicting – social goals? As a tentative step to explore this issue we can note that we are clear on the things we don't want in our societies: crime in the streets, ethnic strife, social division, lack of community spirit, and an uncaring attitude to fellow human beings. This is a start; however we struggle to agree on how to achieve these social goals. And we have serious difficulty in understanding why we cannot agree. There is no more urgent research project than to seek answers to why an intelligent animal with common ancestors and superb abilities to communicate cannot agree on common goals.

The idea of 'sustainability' is to keep things going, but not necessarily the same things in the same way. We have mentioned three aspects of sustainability that we will need to make sense of if our task is to be achieved. What of the 'development' aspect of sustainable development? The idea of 'development' is that of progress, improvement – with the goal of making a better life. This, in the context of the global sustainable development agenda, applies first and foremost to the world's poor – they desperately need a better life. While that is obvious, what does progress mean for the world's rich middle class? That is to ask: what makes us happy and content?

Seen in this light, sustainability requires an understanding of the laws of ecology (to recognise the limits we face and the possibilities open to us), and it also requires agreement on a workable set of social goals within which to allow a market economy to operate. (We can note that not all the goods and services we desire will be provided by market transactions, but most will.) Finally, it asks us to indentify the characteristics of a contented life. One by one we will put some flesh on the bones of these elements of sustainability.

Working with the dynamics of nature

From the very onset it is extremely important to recognise that sustainability is not about 'stopping the world' as it is today; or about restoring it to some pre-existing 'natural' condition. Sustainability is fundamentally about working with nature to provide for human needs on an ongoing basis. Before those who rank nature first stop reading, this does not imply that humans are not part of nature and that the well-being of the other animals with which we share the planet is not a basic issue. There will be a spectrum of views on this matter. Working with nature implies accepting the biological, chemical and physical laws and the processes of evolution that determine the functionality of the natural world. In working with nature (not against it) we are seeking to use an ecosystem's goods and services to maintain us in a healthy and (hopefully) happy state. What do we mean by environmental goods and services? This is terminology borrowed from economics, but its utility is that it makes explicit that nature is of (huge) benefit to us. It will suffice until better terminology comes into being.

We can show schematically how ecosystems provide environmental goods and services. These are the 'flow' resources emanating from the 'stock' of natural capital (that is, the environment). We again are using economic jargon, and please excuse that if it is not to your liking. We are willing to substitute other language if it serves the same purpose. Figure 2.1 illustrates the concept of stocks and flows of environmental services. The ideas encapsulated in this figure are explained in some detail later, but first we need to elaborate on the relationship between the economy and ecology.

Economics and sustainability

The economic reason for seeking sustainability can be stated as follows: a healthy economy depends on a healthy environment. A healthy environment is able to renew its functioning parts (for example, soils are replenished, fish populations are maintained), so that a given population of humans can live indefinitely. Note immediately that the carrying capacity of the environment determines how many humans can live at a given lifestyle. A healthy

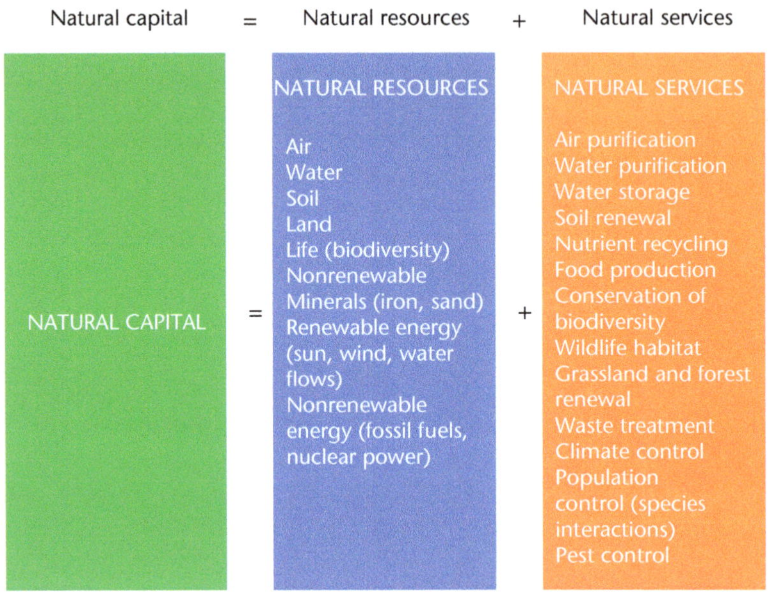

Figure 2.1 Environmental goods and services. Source: Hundloe *et al.* (2011), based on Miller (2007).

economy is resilient. It can suffer shocks, for example the vagaries of nature such as drought and floods and the vagaries of international and domestic markets such as shifts in the exchange rates and changes in consumer preferences (and much more), while continuing to meet human needs and desires. For an economy to be resilient it has to be very flexible, not locked into providing particular products made by particular industries in particular locations. This is one of the hardest things for self-interested humans to contemplate, because there is an understandable tendency for producers of goods and services to attempt to sustain their particular businesses (producing the same products) regardless of changed circumstances. If consumers stop buying a product and if they do so indefinitely, the business will not survive. If you are the producer, you might, if you are politically powerful enough, seek assistance from the government. If assistance is given, it will not save your business in the long term. Only customers – long gone – could have done that.

The short-term self-interest of business people combined with the political imperative of governments to favour the short-term electoral cycle works against the development of a robust, resilient and responsive economy. This is one of the most basic flaws of our political economy. We will not achieve long-term economic sustainability until we devise a better system – that should not take rocket science, just the power of persuasion, for which we need media working for the people, not for sectoral interests.

A big business facing the loss of income (a consequence of declining demand) is likely to approach governments for a subsidy – and be granted one. This, of course, retards the development of a resilient economy. It is ultimately counterproductive. We need to be adamant, given the dynamic nature of the economic world, that it makes little sense to attempt to sustain indefinitely one product, one industry, or a specific industrial region. As mines are depleted, mining towns stagnate and – unless something new can be found to sustain a local economy – they eventually close down. There are many other examples. Often consumer preferences are the catalyst for change. If we had governments sustain the fashion of the 1970s, we would still be wearing bell-bottom jeans! The fact that we might be doing so again in the future with yet another change in fashion simply underpins the need for flexibility and resilience.

What needs to be preserved in an economic sense are the foundations of a genuine competitive market economy. Let us briefly outline how a market economy works – and note, so as not to have a misunderstanding of terminology derail the discussion, that a genuine market economy is not a laissez-faire capitalist economy. The differences can be stark. It is the laissez-faire, not the genuine market economy, that is a major cause of our environmental problems in the twenty-first century. Laissez-faire means that anything goes, including unlimited pollution into local rivers and the atmosphere.

The obvious starting place in economic theory is to state that business people seek to maximise profits and consumers seek to maximise their satisfaction from their consumption choices. In doing their part, businesses need to provide the goods and services consumers are willing to purchase. All other things being equal, the lower the cost the more successful the producer. The low-cost producer will sell more than a high-cost competitor with the same product. The competitor will either find a way to reduce costs or go out of business. The fact that for the vast bulk of everyday goods and services, where there are numerous competing suppliers, prices are the same or close to it, suggests that competition has forced prices down to a low-cost, competitive level.

What drives entrepreneurs to look to the future are 'emerging' profitable opportunities, as Joseph Schumpeter described them. These are what make the economic system dynamic. The 'new' replaces the 'old', and the successful business person will make better-than-normal profits until others wake up and enter the 'new' line of business. Note that economies are dynamic over very short time spans compared to the time periods ecologists and biophysical scientists, such as geologists, deal with. This difference, we will come to appreciate, is one of the two major impediments to linking economics to the environment. The other, to move ahead of ourselves, is the ignorance of consumers – we make far too many choices (use of tobacco products, over-eating, and dangerous driving) that do not coincide with our long-term self-interest.

The dynamic nature of modern economies came to be understood by Joseph Schumpeter. He was one of last century's eminent economists. While much of his political economy is not necessarily appropriate or relevant to a sustainable future, his analysis of the dynamics of an economy is to the point. He used the term 'creative destruction' as an explanation of economic change and growth (see Box 2.1). Old products (which can be relatively new, of only a few years duration), are no longer bought as something more enticing (exciting) appears as a substitute. Some creative destruction leads to progress; some is simply wasteful. The enormous advances in technology and medicine (to take two examples with which we are all familiar) in the past 100 years suggests that great progress has resulted from creative destruction.

The five points in Box 2.1 are the background forces underlying the dynamics of successful economies. In recent decades nothing has been more noticeable than product innovation, particularly in goods using computer technology. Changes are not necessarily for the good – that is a matter for a later discussion. The simple point is that they are happening. Economies are extremely dynamic. Take, for example, the emergence of an enormous middle class economy in China. It is in the order of total population of the United States, and it has formed in only 20 years.

The crucial role of competition must not be overlooked in any analysis of environmental sustainability. An economy will not perform properly (in a textbook sense) in the absence of genuine competition. Monopologists and oligopolists don't have the incentive to provide goods and services at the cheapest possible cost, and they don't have the

Box 2.1: Schumpeter's five categories of innovation

1. New or improved products appear – product innovation.
2. Improved processes are introduced in the production process - process innovation.
3. The emergence of new markets.
4. Discovery of new or alternative supplies of natural resources.
5. Improved management – organisational innovations.

incentive to innovate. This means that economic change and least-cost production are usually only driven by competition in free markets. At this point we will leave this critique of economic theory and practice. We don't want to divert our attention from linking the environment and the economy, which is our real objective.

The various forms of capital

Let us elaborate on the key matters that link the environment to the economy. What are the types of economic values that result from human interaction with the environment? Figure 2.2 contains what economists call 'stock resources', otherwise the various forms of 'capital' that are fundamental in the production of goods and services destined for the marketplace. It shows what goods or services flow – the income streams – from each of the capital stocks.

The environment, or nature if you prefer, is described as 'natural capital'. The other types of capital should be self-evident.[1] In Figure 2.3, the income flows are directly related to their formative capital stock; for example, human capital earns income (wages, salaries, or profit from entrepreneurship). Each type of capital, in conjunction with each other type and with labour, raw materials and the organisational skill of the business person (entrepreneurship), produces a 'flow' of valuable resources. Think of sustainable harvests of wheat or fish. These resources, when sold in the marketplace, earn income for the 'owners' of the stocks of the various capitals used in their harvest; for example, a tractor (human-made capital) operated by a skilled farmer (human capital) on land (natural capital) produces income in the form of the farmer's 'profit'. A fishing boat (human-made capital) operated by a skilled fisher (human capital) at sea (natural capital) produces a conventional income plus resource rent. What we commonly consider to be profit for a farmer comprises rents, interest, profit and wages, and it is the same for an industrialist or other type of business person. Figure 2.3 makes more explicit the nature of the income received.

So far we have said little about the specific 'uses' of nature that generate benefits for us. We are going to need to identify the various uses of the natural features of our case study catchment in a future chapter and here is a good place to sketch the framework.

More detail on the value of nature

Figure 2.4 shows that there are two broad classes of values humans obtain from the environment, known as use and non-use values. There are various sub-categories of each of these values. Whether or not the economic values are recorded by market transactions or are provided outside the market is one defining feature. It is important to recognise that various goods and services that are important for human life and satisfaction are not

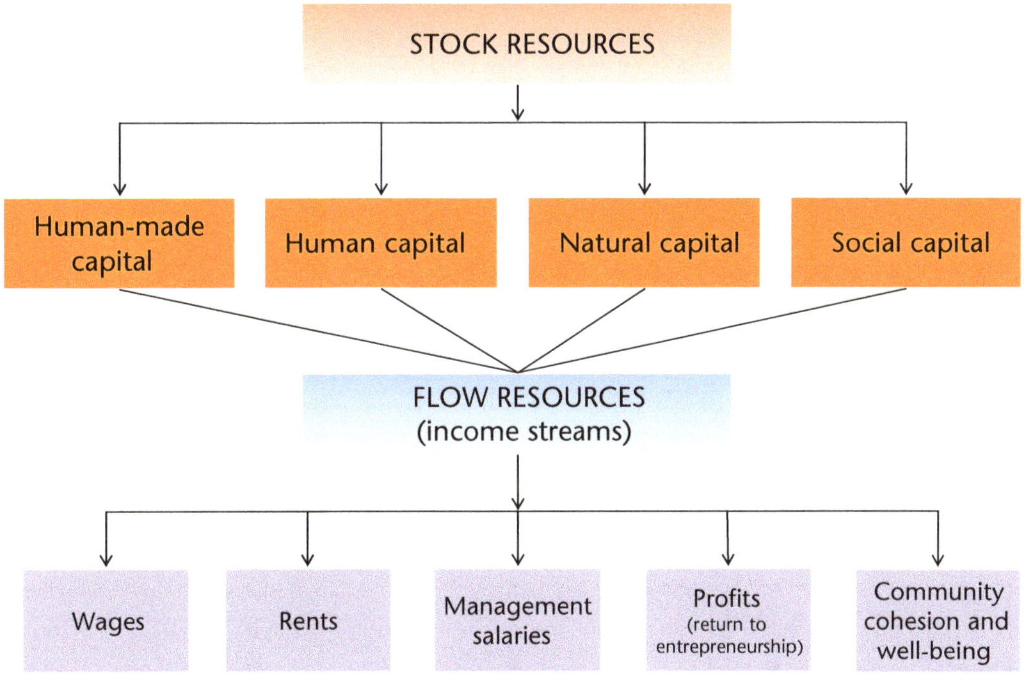

Figure 2.2 The 'capitals' underpinning economic values.

bought and sold in the market, and because of this they tend to be overlooked and undervalued. We can't let this continue to be the case if we are to be rational about human–environment interrelationships and if we are to achieve sustainable development. That noted, we should keep in mind that many non-market goods and services can be valued in economic terms, although with difficulty – and often crudely.

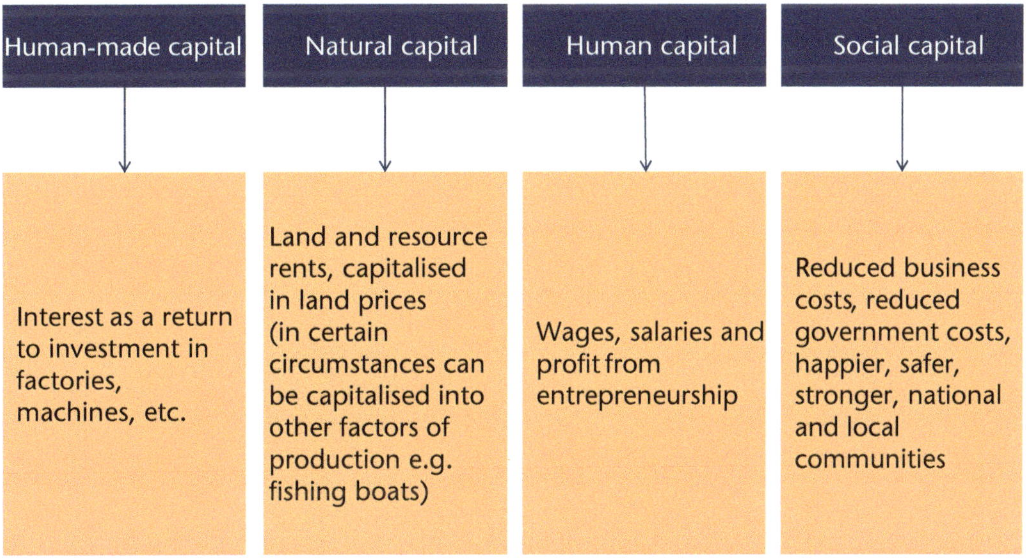

Figure 2.3 What each type of capital earns or saves.

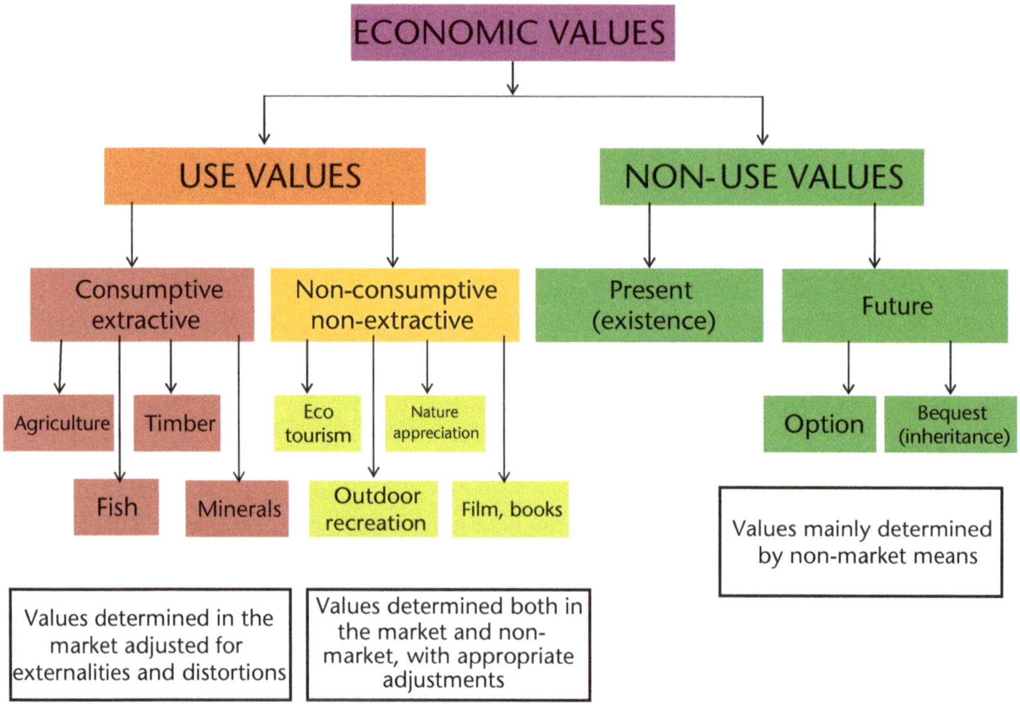

Figure 2.4 Economic values flowing from natural capital.

Society and sustainability

This chapter has so far focused on nature (natural capital) and its role in making for a better, more rewarding life. There is another form of capital that until very recently has been ignored: social capital. Humans are social animals. Much of the enjoyment we get out of life is derived through our social interactions. Families, friends and local communities all play a significant part in our lives. So do special-interest and sporting associations. People who have multiple social links (either formal or informal) tend to be happy, and if others in their community are also actively engaged, the community is likely to be a safe and pleasant place to live – in formal terms, the community is likely to have high (local) social capital.

From a business perspective, social capital is evident when trust is shown by a handshake and the cost of large legal fees is not required. Then there are cases of business people (farmers, fishers and furniture workers) cooperating by lending equipment, helping out in emergencies, and generally supporting each other and the community in which they live. Reciprocity comes naturally when trust, shared visions and community spirit prevail. This situation should not be unusual in a small catchment, particularly if common pursuits are undertaken.

The greater the degree of social capital in a community, the less likely the need for formal policing and the government provision of community services. And the less need for lawyers to settle disputes. In this sense, social capital saves resources; that is, it reduces both business and government costs, and therefore increases productivity. Businesses based in communities with strong social capital are able to benefit from it in various ways:

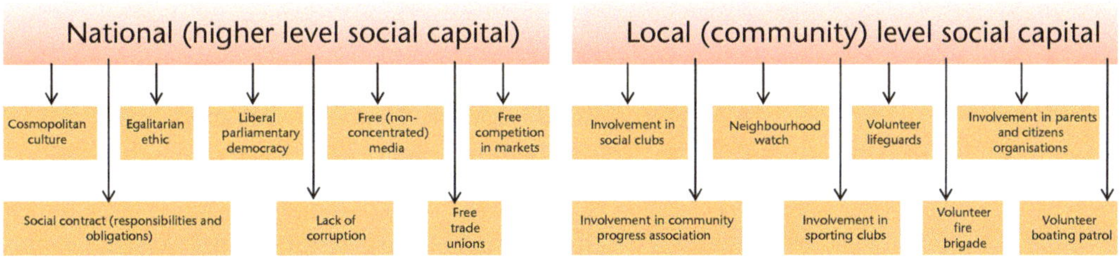

Figure 2.5 Social capital in its various forms.

willing and happy workers, cooperative arrangements with related businesses, no crime, and minimal transaction costs. All social and business relationships are smooth in communities with high social capital.

Social capital is extremely important at the national level. There is a high level of social capital evident in mature societies: a sound system of laws and regulations; a strong financial sector; a corruption-free government; and 'sophisticated' consumers (meaning well-informed). These plus other characteristics of a democratic, stable society are part of a nation's social capital. Strong (and, obviously, sustainable) social capital, plus resilient natural capital and a flexible economy are necessary prerequisites for a sustainable society. Figure 2.5 illustrates the nature and function of social capital at both the national and the local level. The second-level boxes are merely the institutions that underpin a trouble-free social and business life.

Figure 2.6 shows the contributions of social capital to the economy just as we did for natural capital. Of course, social capital does much more than provide economic benefits. It is valued in its own right. It is measurable in economic terms as cost-savings; in other words, the extra taxes or business costs we would have to meet if there was less social capital. The degree of social capital is a parameter of interest in our case study.

To our next step

This brief chapter has presented a framework that we will utilise in future chapters where theory gives way to practice. We will – before getting to the practice – say much more about the precise nature of sustainability (the next chapter) and follow this with a brief discussion on how we can estimate dollar values for non-market goods and services. This is a prerequisite for our case study.

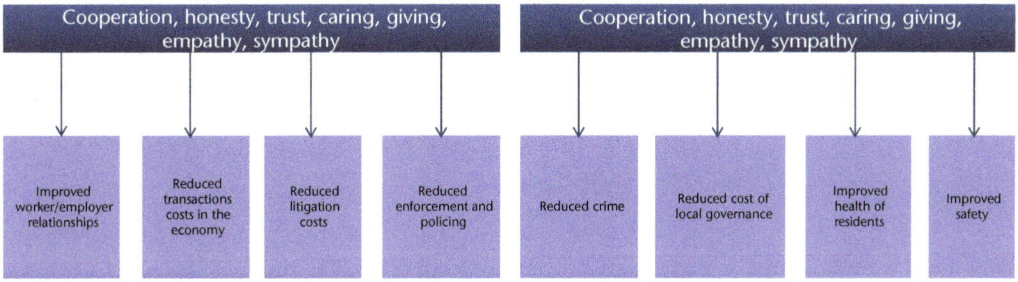

Figure 2.6 Social capital in its various outcomes.

3

The principles of sustainability and economics

Tor Hundloe

We have already pointed out that in 1987 the United Nations' four-year-long investigation of environmental and development matters, the Brundtland Report, came to its conclusions and put the concept of sustainable development on the agenda of governments worldwide. Five principles underpin the concept. Two are ethical constructs, rather than being science-based. The first, and best known, is the principle of intergenerational equity. At its simplest this requires us to manage the globe's ecosystems and economies in a manner whereby future generations will be no worse off than present generations. And for those presently living in poverty, it requires much improvement on their situation. The second ethical principle is a logical complement. It is the intragenerational principle, which requires equitable treatment of all humans today (and in all future generations) – basically the reduction of poverty and the improvement of living standards in poor and developing countries to a level commensurate with the world's middle class.

The other three principles are instrumental (or enabling); that is, they aim to enable the achievement of the ethical principles. There is the protection of biological diversity to ensure the resilience of the globe's ecosystems. Then there is the precautionary principle, which aims to provide for rational risk-aversion when making decisions that could do irreversible harm to the environment (including the social and economic environment). Finally, there is the need to link ecology and economics in decision-making at all levels. It is this last principle that has driven the quest for a sustainability science. The gap between the two sciences of ecology and economics has become a chasm. Much will need to be done to make a sturdy, functional bridge between these two fields of endeavour. We must be realistic and recognise that the task of developing a genuine transdisciplinary approach to environmental problems will be difficult, given that all sciences have become narrower in their focus over the past hundred years. What has been split asunder in the search for ever-deeper specialist knowledge we now wish to put back together in the quest for a much improved understanding of how to manage our way from this day on.

Sustainability and economics

One of the key sciences we are dealing with is more problematic than the others. It is economics. For this reason we need to focus on its good and bad points. There is a requirement to comprehend sufficient economics to make our task of describing and analysing life in a catchment in an integrated framework easy. Those with a background in economics can skip this chapter, yet we argue that economists will appreciate our wide-ranging discussion of their discipline.

A particular problem is the multi-faceted role the discipline of economics plays in our daily lives. Other than trained economists, there is a poor understanding, often a misunderstanding, of what precisely economics is. We tend not to have the same difficulty in understanding the nature of ecology – most of us have a gut feel for what biologists, botanists, zoologists and the like do. While a short theoretical digression might interrupt the flow of our discussion, it is absolutely necessary if we are to make any progress in developing a sustainability science and in making sense of the analyses of our case study catchment, the Little Swanport catchment.

What is economics?

What is economics and economic science? The short answer is that there are two types of economic analysis, and hence the answer depends on what we are discussing. There is what economists call 'positive economics'. This is simply about gathering data, analysing it, and looking for relationships that describe and explain what is actually happening to a particular industry, a national economy, or to households. In this role, economics is a science not dissimilar to physics, biology or ecology. It is value-free, policy neutral, and is used to describe, or model, many everyday economic relationships (including business and consumer decisions). It is used to inform interested parties (more often than not governments) of the state of play in social and economic matters (for example, the level of employment, the level of inflation, or the links between sectors in an economy).

One can analyse and describe a subsistence economy (as still exists in many parts of sub-Saharan Africa and the Pacific), a socialist economy (none warranting this description exists at present), a capitalist economy (the dominant economic form, but having significant variations) and anything in between without coming to any judgment about their desirability. This is what economists as trained technicians do. It is extremely important work in a modern society – yet its very neutrality limits its usefulness. It does not inform us of the pros and cons of the particular economic structure and relationships. There is no yardstick against which to judge the economy under scrutiny, or an economic policy a government is contemplating.

In contrast, there is 'normative economics'. The adjective suggests that this is a economics that is directed towards achieving some (presumably desirable) goal. The easiest way to describe normative economics is to refer to a perfectly competitive economy as the yardstick. Keep in mind that this particular economic system is not the only one that can be society's preferred goal. It has become the yardstick for the sole reason that no other one is presently being advanced to take its place. There are no competing models on the horizon, and hence we will stick with it as the yardstick until there is reason to change. This stated, it is important to note that the ideal (model) of perfect competition is not all black and white.

The competitive economy

Perfect competition in its pure form is said to deliver a number of desired benefits. Let us, ever so briefly, describe this form of economics. The non-economist is likely to discover that the ideal model is far removed from the popular view of the economy he or she lives in. Our description can start with us either as consumers or as producers. Most of us play both roles – certainly all are consumers.

Here are the rules. First, producers, of whom there must be many (and hence with no market power), will compete fiercely to provide you and me as consumers with the goods and services we are willing and able to pay for. They will do this in search of their goal of maximum profits – if not maximum, at the very least 'satisfactory' profits. Fierce competition between the many producers will keep prices to a minimum. As consumers, we are in the crucial position of determining what will be produced. We set out to maximise our satisfaction ('utility', in the economic jargon). As the economics textbooks state, 'the consumer is sovereign'. If the producers do not make or grow what we are willing to pay for, they won't sell their products and won't last in the business. If an innovative producer gets the jump on competitors with a new product, he or she will make super-normal profits until the competitors respond and compete away the excess profits. The social outcome of this (perfectly competitive) economic system is the production of what you and I want – and can pay for – at least cost. It is hard to argue against this. Cumbersome bureaucratic management of production lacks the flexibility of a fiercely competitive market. Conventional socialist economics failed for this reason, while the theoretical idea of a market economy embedded in a socialist framework has only been written about, and not yet put into practice.

There are numerous other prerequisites for a competitive economy. For example, not only are producers to have no power to influence markets and prices, but the same applies to buyers – a gigantic supermarket that is the only purchaser of farmers' vegetables and fruits has too much market power. The farmers will be severely disadvantaged and obtain low prices. This is what is termed monopoly (one seller) or monopsony (one buyer). If there is more than one, but only a few (as is the experience in the major retail trade in Australia), we have oligopoly on the sales side and oligopsony on the purchasing side. These forms of market structure break the rules of a competitive economy.

Another fundamental prerequisite of perfect competition is the requirement that consumers and producers have something approaching perfect knowledge. If they do not, this is not a critical flaw when the goods and services in question are frequently bought and sold – in this situation, over time all parties acquire knowledge of a product's quality, reliability and value for money. Loaves of bread, haircuts and petrol are consumer purchases where the buyers are likely to be very knowledgeable. But when it comes to new 'high tech' products (here we should include a range of 'e-products'), and the mysterious doings of nature (e.g. worms in the soil), we have great trouble making informed decisions, because we lack the required knowledge. We tend to place ourselves in the hands of (very) young salespeople when we choose a computer or mobile telephone, and we tend not to do enough digging (pun intended) when purchasing a farm – how good are the soils, are the worms doing their job? The requirement of perfectly informed market participants would condemn economics to a category of useless disciplines if we were not willing to override it. How far removed from the perfect can we be allowed to step before the utility of the model can no longer be defended? This is still being debated.

To the next difficulty with the model, of great concern to us in this era of population pressure, the threats of climate change and extreme climatic events, is the exclusion of the costs of pollution, environmental degradation and resource over-use. Only by the recourse to law can we make our economic system account for these negative features of our modern lifestyle. There are things we value but they are not bought and sold in markets; the most obvious of these are clean air and clean water. A large number of unpriced environmental goods and services need be considered in any assessment of how best to live and what is best to do for the future of the planet. Much of this book deals with these environmental values, because we tend to find them in greater numbers in river basins than in grasslands or deserts. Unless we can bring these 'externalities' (unpriced goods, services and actions) into our economic analysis our only option is to deal with them under a legal and bureaucratic framework.

Correcting market failure

What we have been discussing is what economists call 'market failure'. Much applied normative economics focuses on 'correcting' these failures (oversights, if you like) of markets. The fact that no one pays directly for the cost of pollution in a product's price is market failure. And because this has been happening from the time of the first market transaction, we have to reverse a history, going back millennia, and this requires a significant cultural shift. We are asking for a radical change in behaviour. It is a fundamental (although unstated) part of our culture not to pay for pollution – even though it forces costs on to our fellow human beings.

Otherwise rational people bury their heads in the sand and do not want to accept that they should pay a pollution tax. This is notwithstanding the case that with pollution some people (and other animals) pay through ill health, even death. Only if we are lucky is the cost of cleaning up the pollution the only cost we bear. The costs of pollution are 'external' to the producer and consumer; hence they are called 'externalities'. Uncosted effects have to be valued and included in the supermarket price of goods, otherwise prices are not telling the truth. Unpriced, yet valuable, environmental goods and services are wasted – even destroyed – as a consequence. Beautiful views are destroyed because no one owns them and home builders fight for the front spot. Where access to common property is free, resources such as seafood are over-harvested. We know this is silly, yet our market-determined price signals neglect reality.

One reason for the existence of these socially and environmentally deleterious situations – one that is pointed out by economists – is that no one person (or business) has the property right to clean air, clean water, a beautiful view, and any of the other environmental attributes we seek. There is truth in this statement, but it does not logically follow that we can remedy the situation by attempting to create property rights in parcels of circulating air or ever-moving streams. We have more scope to assign property rights to valuable terrestrial ecosystems; however, there is no need to do this as the government can act to maximise environmental benefits while retaining public ownership. Thankfully, there are many means of solving the problems of externalities and open access.

Is the correction of market failures all we ask of economists? If all market imperfections were eradicated, some (many? most?) economists assert that there would be nothing left over for the economist, or governments for that matter, to do. Economists who take this line of reasoning are called 'economic fundamentalists' or 'free-market libertarians', meaning that 'the market', not politics, should govern all social and economic interactions. The only substantive goal these free-market economists seek is to see produced those goods and services consumers are willing to pay for. This is what they mean by 'economic efficiency'. The willingness and ability to pay determines what is produced, when and by whom, using whatever factors of production are most efficient to use – again determined by markets for these factors of production (labour, land and capital).

One might expect that there are economists who have different (broader) goals in mind for society – in particular goals that aim for equity – a sharing of the benefits of economic life – not simply economic efficiency. And so there are. We will need to consider these more broad-minded economists when we describe the social, economic and environmental relationships involved in living in our catchment economy.

More than efficiency

The non-fundamentalist economists, who are both pragmatic and attuned to social needs as much as private demands, seek a range of benefits and not just economic efficiency from an economic system. For example, they argue that human relationships and human well-being should not be determined solely by market transactions. They believe that personal relationships are not to be bought and sold. They understand that if things go wrong, you need to be wealthy to be healthy. This is morally unacceptable. These economists are likely to have goals such as full employment, fair rewards for labour, and an equitable distribution of wealth and income in society. They also seek a range of positive environmental outcomes. These economists are the ones who can bridge the gap between ecology and economics and further the goals of sustainability science.

Economists of the non-fundamentalist bent would say that constraints have to be applied to the goal of economic efficiency. That is, only after higher-order social goals such as full employment and environmental protection are guaranteed should we seek economic efficiency – not the other way around. This is one of the most crucial arguments to be made in our search for sustainability. Be ever mindful that economic efficiency is not to be the dominant social objective, because without a human face economics unbridled diminishes the human spirit, degrades social relationships, and damages the environment. None of these outcomes makes for a happy human existence. The case for putting 'the economy' in its best place, rather than a domineering place, logically follows. In a nutshell, the market has to be overridden when the outcome is not socially or environmentally desirable. Environmental economists would call this approach 'constrained optimality' – optimal, but within constraints.

The economic jargon used here might be boring, but it is extremely important that we become familiar with the language of economics (and of course, the languages of ecology and ethics) if we are to have meaningful discussions about environmental and sustainability matters. Too many unnecessary disputes occur because of a misunderstanding of language, not of ideas. When we all speak the same language we tend to find that we seek the same goals. Let us hope so.

The future compared to the present

In searching for a sustainable future based on the marriage of economics and ecology, there are more problems to be resolved than the neglect of pollution and the disregard for common property. An obvious case where conventional economics needs to be overridden in the endeavour to achieve sustainability is the procedure of discounting future streams of costs and benefits. Discounting is what economists and engineers do to convert costs and benefits that are in the future to today's prices (the net present value). This has nothing to do with inflation.[1] It is purely about valuing the future according to a set of economic rules – rules, not ethics. How we feel about the future – in particular, about the way future generations should live and about the future of the planet itself – is not a consideration in standard approaches to discounting.

In the conventional process of evaluating projects (say, a large water storage) where the costs and benefits extend far into the future, the year-by-year monetary flows (expenditure or income) are converted to a net present value by the use of a discount rate. If sensitivity to the discount rate is considered important information by decision-makers, a range of discount rates, not just a preferred one, will be used for comparative purposes. However, if the project is a government one or one requiring government approval, the public is not necessarily kept informed on the discount rate, or rates, used by governments. We are seldom told which discount rate is the 'swing' one, that is, the one that determines if there is a benefit to cost ratio greater than one. It seems that society's time preference rate is for bureaucrats or politicians to decide.

Box 3.1: The dramatic diminished value of a dollar

To obtain an understanding of the harsh impact of even small discount rates, keep in mind that in 50 years a 10 per cent discount rate would reduce the monetary value of $100 to about 80 cents in today's value. Non-economists and those not attuned to the mathematics involved should read this again. A dollar today is worth 0.8 cents 50 years from now.

Assuming only one rate is considered, it could either be the market-determined price of borrowed money (the opportunity cost – given the money could have been used for other investments that we would expect to earn a positive income) or a generally lower 'social discount rate' set by government treasury departments. There are other options for determining the appropriate discount rate, but these two are the conventional ones; however, none meets the criteria of intragenerational equity, treating the future on the same footing as the present unless the 'social time preference rate' equals a zero discount rate. We should note that eminent ex-World Bank economist and climate change economic analyst Lord Nicholas Stern uses a discount rate very close to zero in assessing the costs and benefits of climate change. To understand the dramatic impact of discounting see Box 3.1.

Anything other than a zero discount rate is most likely to conflict with the intergenerational equity principle. We could say 'will conflict' rather than 'most likely', but we use this specific terminology because there is the possibility that we might need to favour the present generation over the next one! This would be the intergenerational rule if we expected that future generations were going to be much better off than we are. Obviously, we would need to be very careful in how we came to this judgment. That accepted, we need to recall that our grandparents worked extremely long, hard hours and had a much poorer lifestyle than we do – so that we could be better off. That is not equitable, but we cannot compensate them. We do not want to make this mistake again. While our forebears were likely to have been driven by personal goodwill, we have a number of examples, the Soviet Union and Maoist China, where a generation (or more) was deprived of consumer comforts in the name of rapid industrialisation, which was going to better future generations. The collapse of 'communism' can, in part, be attributed to the loss of support for this intergenerational inequity.

The convention of applying a discount rate to evaluate the future flow of costs and benefits is so ingrained in economics (and financial accounting) that environmental economists working in transdisciplinary sustainability science simply have to set their own rule – do not discount – or relegate conventional cost–benefit analyses and project appraisal approaches to the waste-paper bin. Conventional cost–benefit analysis is not appropriate for the analysis of a project's sustainability.

Redoing a conventional cost–benefit analysis using a zero rate is preferred by many practitioners. A zero rate is neutral in the weighting of the future. Applying a zero discount rate will have dramatic effects on the assessment of projects that have costs far into the future; for example the gradual silting up of a dam and the need to rebuild it will hardly register as a cost in a discounted project evaluation, yet it is going to be a substantial cost when a zero discount rate is used. Conversely, benefits that accrue well into the future will also be so small as to be disregarded in a conventional analysis but will be significant with a zero rate.

Income distribution in valuing projects and policies

We now turn to another matter that requires attention in our endeavour to integrate economics and ecology. It relates to another matter of fairness. Here again, conventional cost-benefit analysis fails us, because its focus is purely and simply economic efficiency. If we are to undertake economic analyses that are consistent with sustainability principles there is a need to account for the highly unequal distribution of incomes we witness throughout the world, and within nations. Recall the intragenerational equity principle, which states that we should be favouring the less-well-off – and ultimately treating all equally. Projects that favor the rich, without in some way compensating the poor, are, on the face of it, at odds with sustainability ethical principles. This is not a major problem in the rich countries of the world. They are middle-class societies with a very small group of very rich and another small group of poor, and it is usual for the government to attempt to assist the poor, by welfare payments and other policies, although the attempt often fails. A blind eye is turned to the super rich, notwithstanding serious questions that could be asked about the means they have relied on to attain their status. Hard work, brilliance, or both, are not likely to feature. The 'desert' theory – they deserve their riches – is not a serious proposition. This brings us to a fundamental but neglected economic truth: 'a dollar is not a dollar' until we know who receives it. Read on and we will explain.

How do we untangle the starkly stated proposition that a dollar is not a dollar? In the first instance, we need to assess whether a project or policy favours the rich or the poor according to an empirical test. And, of course, we have to decide where to draw the line between rich and poor. In middle-class societies with generous social welfare systems, such as the majority of OECD countries, it is only if a project favours the very rich (who are small in number) that we would need to consider its effect on intragenerational equity. The situation is different in most of the emerging and poor countries, where the very poor are numerous.

It is possible to value a project or proposal to represent the relative value of additional benefits and additional costs to either the poor or the rich. First, we need to establish the theoretical economic justification for adjusting monetary flows according to the wealth and income status of the affected party. Start with the fact that an extra dollar is worth less to a rich person than a poor person. This is true, is it not! Economists have observed the fact that there are diminishing benefits to a person as more money is obtained by that person (rich or poor), just as there are for ice-creams eaten one after the other. All normal goods and services fit this rule. This is the law of diminishing marginal utility. The first ice-cream on a hot day is valued much more highly than the second, and the second more highly valued than the third. We would be willing to pay considerably more for the first than the third. Our satisfaction diminishes as we eat consecutive ice-creams over a set time, and so does our willingness to pay.

Is this economic fact simply too radical to be spoken about? Surely not! However, when it comes to the application of diminishing marginal utility of income (money or wealth) it is not mentioned, and it is very seldom recongised in assessing projects or policies. It is the rare financial analysis or economic assessment that takes this into account. We will need to rectify this if sustainability principles are to be met – for the simple reason that equity matters as much as efficiency. Economists should apply the principles of diminishing marginal utility to income in the assessment of policies. If the principle were applied, it would have an impact on what projects were funded.

Take another example, the reason why workers have to be paid more per hour ('overtime') to work longer than standard hours, or to work when others are on holidays, is that workers value the extra income less than the income to be gained from the previous hours (or days) of work. An incentive in the form of higher wages has to be paid if they are to be willing to work additional hours (of course, the decreased leisure time is becoming more valuable the less of it they have). A dollar is not a dollar!

A dollar is not a dollar

If there were not diminishing marginal value we would not have a normal demand function (or curve) and the law of demand and supply – the one law of economics everyone understands – would not apply! This is how fundamental diminishing marginal utility is; it is the very basis of economic science. Why is such a basic economic law neglected when the matter is the value of extra income? Let us see.

Some will trot out all sorts of furphies about the impossibility of making interpersonal comparisons of utility as an excuse to disregard the law of diminishing marginal utility. If we cannot compare the relative degree of satisfaction of a poor person to a rich person, what is the point of economics? Practical people – non-economists – know that we can. If you think about it, the concept of taking equity into account in economic relationships is common place. Democratic governments of all political persuasions make adjustments to the unequal distribution of income and (sometimes wealth) within society. They do this by applying progressive income taxes, and by providing a 'social wage' or subsidies to those unable to earn an income. The voting public accepts this as a key component of the 'social contract' binding citizens to each other, and binding voters to their elected representatives. In this day and age, grossly unequal treatment of citizens would probably see the defeat of governments at the next election.

If modern democratic governments did not embrace a notion of equity there would be poll taxes (one set rate of income tax) and no social welfare payments. No modern government neglects equity principles to this extent. Thinking beyond the nation state, there are attempts to modify the extreme income differences at a global scale by the provision of development aid, debt forgiveness, and technology transfers from the rich countries to the poor. These interventions are a recognition of the diminishing value of money as much as some moral imperative to help the poor. What we the rich give away is of less value to us than it is to the poor recipient.

The importance of equity issues cannot be overstated in economic analysis. Fairness exercises the mind of most people, well before the arcane matter of economic efficiency. We take it that this point has been made strongly and shall proceed. In much that follows, particularly where monetary amounts are involved, we will be taking market-determined values on face value – no attempt is made to adjust according to equity criteria. As stated previously, this is not a terribly serious problem in the rich country such as Australia. Our case study will not be affected by the relative wealth or income of the categories of people living in the catchment. However, it is open to someone to argue the case that the law ought to be applied even in a wealthy country like Australia. That noted, it is an economic fact of life that who gets what is determined by who owns what, or who earns what. Put in the formal language of economics, willingness to pay is constrained by budgets; that is, by our ability to pay.

4

Economic values of nature

Tor Hundloe

In this and the next chapter, we take up the challenge of how we might formally link the economy to the environment. First some background which follows on from our previous discussion. It has become increasingly obvious during the past 50 years that nature is being degraded through the release of pollutants and widespread over-use of natural resources, both finite and renewable. It costs polluters nothing to use and abuse the rivers, the oceans, the land, and the atmosphere. Thor Heyerdahl, in the middle of the vast Pacific Ocean more than 60 years ago, was astounded to find large blobs of pollutants drifting hither and thither. No one owned the oceans and could sue the polluters for the damage caused. Very few saw the pollution – Heyerdahl and his companions were on the tiny raft *Kon-Tiki* drifting with the currents, and could see what ship captains couldn't from their elevated wheel-houses.

Once pollution was recognised to be widespread and in places expected to be pristine, the call came for economists to value nature in monetary terms – not just a particular threatened ecosystem but the whole shebang. The argument was – and is – that unless nature is put on the same footing with marketed goods and services, the former will continue to be seriously undervalued and hence degraded. After decades of doubt, the non-economists concerned about the future of the environment came to accept that it was better to join 'the enemy' than to fight a losing battle. The enemy was, of course, economics. To link economics and ecology was to prove no easy matter as we will see. We will come to ask: is the marriage really possible?

A heroic attempt

A heroic attempt was made to 'value nature' in 1997, when Robert Costanza and a near football-team of economist colleagues published a study in the journal *Nature* that purported to value virtually all of nature – the globe (Costanza *et al.* 1997). Their analysis showed that the globe as a natural resource was worth a multiple of the total GDP generated in the world. Nature lovers and ecologists were overjoyed. Their enemy had become nature's saviour. While this work grabbed the attention of ecologists and environmentalists (and naturally the media), most economists did not take a great deal of notice because they recognised the limitations in the study. Allow me to explain the reservations of economists by way of analogies.

The money we are asked to pay in the shops (for apples, oranges, coffee and cake or whatever) is the result of vast numbers of people dividing their individual budgets up between the multitude of goods and services available. For example, some or all might forgo cake with their coffee for many years so as to accumulate a sum large enough to purchase the latest television set. If this is considered too farfetched, they will certainly consider the price of apples in relation to those of pears in deciding which ones to purchase. Small purchases – made at the margin, an extra one of this or that – are the crucial determinants of the price of anything. By and large, our decisions to purchase goods and services are small, incremental decisions, considering what is forgone if *A* is purchased rather than *B*. We will explain the importance of this 'marginalist' approach to measurement of economic values.

You might, quite correctly, point out that some of the most important economic decisions we make involve large sums of money. Certainly the decision to purchase a new car is significant in terms of the money spent. However, we value the car's service, not the car itself, in comparison to other equally significant items we could have purchased. If we purchase the car on a time-payment scheme (the way most people purchase a house), we are obviously thinking in terms of relatively small amounts paid per week. In the economics of the marketplace we are dealing with small sums for small purchases. We are forced by a limited budget to make comparisons 'at the margin' and to take into account the value to us of other things we could purchase with the same amount of money – a new car or a deck on the house.

The globe is not a small item. We cannot – as much as we might wish – buy another one, or two! Yet the *Nature* study suggested (notwithstanding the important caveats made by the article's authors) that the planet was being valued as we would in a one-off sale. We could stretch the imagination and assume a team of Martians loaded with cash landed and made an offer for the planet. Of course, very small parts of the planet are sold on a one-off basis. Tiny blocks of land (with buildings on them) are bought and sold daily in New York, at very high prices per allotment. But try to find a sole buyer for the place we call New York City! Keep in mind how the monetary value of anything is a function of supply and demand. If tomorrow every apartment and house in New York was offered for sale, one would expect the average price to drop dramatically. Immediate over supply would drive the price down.

Let us assume that everyone living today owned a part of Earth. Some would own large parts,[1] others tiny parts, and in reality, many would own no part at all. Let us assume that the majority of us would like the cash that the sale of the total globe would bring if one or a very small group of us bought it from the rest. We would expect that there would be billions of sellers and very few potential buyers – until the price fell far enough to entice buyers into the market for the property described in real estate advertisements as 'Earth'. If selling the planet is too far-fetched to contemplate, consider selling a country or a large city.

To put a price on a city or a country using market (or surrogate market) prices is not possible. If this is the case, valuing the whole world is even less practical. Economists understand that there is no possibility of finding one buyer for the globe. What the authors of the *Nature* study did not emphasise enough (there was some reference to it) is that market values (economic values) only make sense as small changes in demand and supply occur – not as one-off sales of an entire city, country, or globe. We value things in economic terms in relation to each other – apples compared with oranges bought in a competitive market, visits to national parks compared with visits to museums. To talk of the world's measured GDP as the economic 'value' of the world is nonsense.

There is another problem. Using GDP as a measure of economic value is itself a problem. GDP is nothing more than a monetary accounting measure of goods and services bought and sold in one year. If they are bought they must be valuable! Not so. You might think that we value – that is, actually desire – all the things produced and purchased in the market. Not so. Consider the expense in repairing cars damaged in accidents, and the costs of repairing humans who were in the cars when they were damaged in accidents, and costs of employing an additional person while the person injured in the accident is in hospital recovering. All these outlays of money go into the derivation of GDP, as do the costs of cleaning up pollution such as the litter thrown from cars, and the pollutants, including the greenhouse gas carbon dioxide, emitted into the atmosphere. None of these things did we want. The money spent on repairing damage is only necessary because we did not have the good sense to avoid the damage in the first place. The more damage done the greater the GDP! In adding these costs to a measure that is supposed to represent benefits, we have a serious problem in measurement. Then there are the truly valuable things omitted from the calculation of GDP. All the work done in the family house and garden does not get included, unless one hires a housekeeper and gardener. Marry your gardener or housekeeper and GDP declines! This is serious. Our national statistics don't lie. GDP is a very partial measure, and even if it were sensible to measure the globe as a natural, functioning ecosystem, its comparison to the world's GDP would be misleading in the extreme. That is the point. The lesson is: don't attempt the impossible.

Don't attempt the impossible, do the practical

In the later chapters of this book, when we come to link the economy of Little Swanport catchment to its ecology, we are forced to revisit these issues of economic valuation. However, we have a useful (and theoretically valued) solution. What linking economics to ecology means in practice can be expressed in a simple question, not the unrealistic 'what is the economic value of the Earth?'. Here is an example of a relevant simple question: if we returned a little bit of water in a catchment to increase environmental flows, what are the economic benefits of that action compared with its economic costs? Some users, say farmers or the operators of a hydro-electric power station, would lose some water, which would cost them in reduced production and income. Their losses or costs need to be compared with the monetary value of a healthier downstream environment and the gains in income of downstream water users. This is the way economists consider and measure value, whether it is water in catchments, animals in native forests, fish in the sea, or any other environmental attribute we are interested in. In economics, as practised today, values (prices) are relative to each other. This allows economics to function as the science of choice. It is unique in this regard, but as stated a number of times it has limitations. In ethics and other human pursuits there are absolutes, and this distinguishes them from economics.

Let us continue with water as a free gift of nature that we want to value in economic terms. Ideally an economist would like to do an experiment that would involve incrementally altering allocations of water (in small doses as a medical doctor might do with drugs) until the benefits of the last unit to the environment or a downstream water user just outweighed the costs of the last forgone unit to upstream water users. It should be obvious that the reallocation of water could be moving in the other direction – allowing a greater take upstream rather than to downstream users. It would be the same experiment. At the stage where there are equal benefits per unit of water upstream and downstream the economist would call a halt to shifting water between users, as no additional incremental gains could be had whether upstream or downstream. This is the strength of marginal analysis in economics.

In practice there are very limited opportunities to undertake an experiment like this. The real world of farms, forests and fisheries is not a scientist's laboratory. This differentiates economics and other social sciences from those sciences where theories can be tested in the laboratory, without impacting on the well-being of another human being. What an economist would suggest as a real-world, practical experiment is to establish a market that allowed water to be purchased by downstream and upstream users. Economists would expect that, given enough time for buyers and sellers to experiment (that is, to evaluate the results of different quantities of water), water would come to be purchased and used in such a manner that no possible economic gain would result from a further reallocation at the margin.

The neatness of the economic-preferred solution of a free market for water is only available in very limited situations, as governments are hesitant to allow a genuine market to operate. The history of water 'rights' and water use imposes practical constraints. The politics are too difficult. The economically powerful (such as power-station operators and large-scale manufacturers) and the weight of numbers in the cities would determine who came to get water. We would expect that the majority of farmers, individually and collectively lacking market power, would be the losers. Recall that city folk are used to paying much, much larger amounts for water than farmers and industrial water users. If it came to a bidding war for water in a severe drought, the economic wealth and incomes in cities would result in their needs and wants being met, notwithstanding the fact that animals were dying and crops withering away on farms. You would be correct to point out that if city people understood their long-term interest (they purchase farm products) they would not use their market power to destroy farming unintentionally. The fact of the matter is that most people do not make this sort of connection and would only come to realise the counterproductive effect of their behaviour too late.

The above discussion suggests that we should not expect a sensible answer to questions such as what is the monetary value of the Nile River and Delta. However we could answer other questions of an economic bent. For example, what is the monetary value of the products grown along the Nile flood plains? We could, if necessary in making a water allocation decision, compare this value to the loss of denying the downstream Nile its present water flow. However, this measure would be of a completely different type of measure to that of the environmental value of the river. We would not know, without considerable additional analysis, what the actual contribution of the river water is to the production of what is grown (given that the produce grown requires much labour, the use of considerable capital equipment, and the application of fertilisers). Water is only one input, although an essential one. We could attempt to estimate the value of additional units of key inputs by formulating a production function (see later for a discussion of production functions); however this exercise is very demanding of data and the results often prove inconclusive due to the variability of farming practices and prices of farm products. While it is possible to formulate a production function for an individual farm, it is not easy to generalise about the average farm unless there are similar farms (in size, in crops grown, in inputs used) in a particular location. There usually are not. This is the reason that economists do not know the value of water (as a farming input) in the world's catchments. Only by case-by-case analyses can we provide useful answers.

If we find it extremely difficult to construct production functions based on analyses of individual farm experiences, there are other opportunities to achieve our goal of obtaining the value of a little more or a little less water to the farmer, fisher or furniture-maker (all need water). What we need to do to obtain the value of an extra unit of water is to observe the effects on (farm) income of a unit reduction or a unit increase of water where nature takes care of this experiment by varying rainfall. In catchments where there are large water

Box 4.1: The components of total economic value

Total economic value = Use value + Non-use value

- Where *use value* = the monetary value of goods that humans use. These can be either *final consumption* goods, such as firewood taken from a forest, fish from an estuary, or berries from a forest, or *intermediate goods* such as the agricultural soils in flood plains that have been converted to farms, or the water extracted and used for irrigation.
- Where *non-use value* = the monetary value of services that humans benefit from but do not *use up*(do not *take)* and hence do not deplete or degrade. These include the value of a nature walk, bird-watching, or experiencing a beautiful view, and much more.

How to measure the various types of value

- *Use values* are generally observable in the prices paid in the *market*, such as the purchase of good quality farm land; or *would* be paid in the market if one existed, such as a right to go hunting on a privately owned land, or take dead firewood from a forest.
- Some *non-use* values are observable in markets, such as the premium paid by a house buyer seeking a beautiful coastal or mountain view, or when a tour operator is involved in taking groups on treks for nature appreciation.
- Other *non-use* values (where no market exists) have to be measured by the use of techniques based on either surrogate or hypothetical markets. An example of a surrogate market is where the consumer pays travel expenses and uses time to get to appreciate a desired experience such as a visit to a national park. A hypothetical market is where, by the use of a questionnaire, a respondent gets to nominate his or her willingness to pay for, say, the preservation of a threatened, beautiful site.

And yet other values

There are other categories of economic value that are not covered in the above.

- First there is *option value*, which is willingness to pay today to ensure protection of the natural environment that could become of value in the future. The protection of rainforests – an excellent example being the Amazon River Basin forest – is thought to be wise, given they are considered to hold enormous potential in providing (yet to be discovered) pharmaceuticals. If we, as potential consumers of the drugs, expect good results we might be willing to pay more for protection than, say, a woodchipping company would to access the same forest (and destroy its drug-production potential).
- Second is *inheritance value*, otherwise known as *bequest value*. This is our willingness to pay today to ensure that the environment in question stays intact so that it is available for the appreciation of future generations. Again our aggregate willingness to pay has to be large enough to outbid another potential uses (for example, someone wishing to destroy the environment for immediate financial gain).
- Finally there is *intrinsic value*. This is when people pay to save an environment for no other reason than they want to save it. They may never plan to visit it or use it. This type of value has to be estimated by the use of techniques based on hypothetical markets.

storages, humans could get into the act by controlling the water flow from dams. Release less water into the system and we will be able to calculate the value of any reduction of (agricultural) production. Production lost, what economists call a 'damage cost', resulting from some adverse environmental circumstance, is often the easiest approach to obtaining a valid value for marginal units of water.[2]

Doing what we can

Water and the catchment it flows through provide benefits to humans over and above the essential amount of water required for survival. These other benefits are extremely valuable; the greater the extent that we can document and measure them, the better. This is not to suggest that we should seek to put an economic price on all of the things that relate to a catchment.[3] Putting that aside, we should be able to measure many of the benefits of the ecosystem goods and services of a river basin, for example the willingness-to-pay for a day's outing fishing in the estuary. Economists talk about 'total economic value' when they wish to include the total range of benefits obtained from a natural environment. The various components of total economic value are described in Box 4.1.

Considerable expertise has built up over the past 40 to 50 years in developing and refining valuation techniques for non-market goods and services. All depend on obtaining a substantial amount of data. For example, the use of the hedonic pricing method in an attempt to value beautiful views experienced by home owners requires a large number of sales of similar houses, except each has a different view. These conditions are difficult to meet, even in large cities. In low-density, semi-rural natural environment settings it has proven extremely difficult to apply this method due to insufficient sales in a comparable timeframe.

The use of contingent valuation and/or choice modelling as tools to obtain non-market values requires relatively large samples of respondents, and the most reliable data is gathered in face-to-face interviewing, a very costly business. There are numerous issues to be resolved using this technique. The researcher is forced to assume the strict rationality of the respondent. This requires the interviewee to have a certain amount of knowledge about what is being valued and be willing to make a realistic trade-off in terms of money for the ecological attribute being valued. The most serious reservation with using hypothetical market measures has to be that respondents do not comprehend the ecological relationships and functions that determine the environment's actual value. Most people only have a vague understanding of ecology and consequently can over- or undervalue certain relationships. Many people don't appreciate the value until it is actually lost. These are serious problems in obtaining economic values.

5

Practical measurements: water as an ecosystem good

Tor Hundloe

We started with the proposition that catchments (or river basins) would form ideal national boundaries. Although a river can wind through myriad sub-ecosystems (from mountains, to grasslands, to wetlands), it is the life-blood of the system, falling water, running water, underground water and evaporating water, that captures our attention and presents both the opportunities and constraints that define our natural economy. If a catchment (presumably a large one) was a nation, the first thing we would do is construct a comprehensive model of its economy. Our treasury departments, in particular, and our governments in general, could not operate without a very good model of their economy. We will go to this matter immediately; however, we draw attention to Box 5.1, to remind us where our analysis is headed.

Box 5.1: Where we are headed

Because our conventional model of national economies is limited to the flow of goods and services exchanged in the market and hence neglects very important values, we will need to attempt to incorporate the value of both environmental goods and services, and the value of social capital, in our description of the economy we are analysing. We have discussed these issues in some detail above. The point is made here to remind us that when we use conventional economic measures to understand our case study catchment there are additional values that are not measured. For example, there are free products such as fish, and there are community relationships (social capital) to consider. This means our catchment's real value (or 'regional product') is greater than our input–output tables and (flow-on) multipliers indicate. To account for this, we will need to estimate the monetary value of the non-market goods and services.

The other matter of great interest to us is the value of water in various uses. We have had a little to say about this above, but when we come to our case study we will need to do our best to deal with this matter.

A catchment as a hypothetical country

At this stage the non-economist reader deserves a warning. To gain a working understanding of the economic life of our catchment we are going to need to use some jargon. However, every time we do, we will explain the meaning of the terms by example. If using jargon is not a serious enough challenge for non-economist readers, we are going to do something that, at first impression, might seem peculiar. In this chapter we are going to think of a catchment as we would a country.

Countries import and export goods and services as a general rule, because they are not self-sufficient and value the benefits of trade. No one country is able or, if able, willing to produce everything its citizens desire. And so it is with catchments. As we will discover in a future chapter, some of the world's largest catchments produce considerable wealth through exporting foodstuffs. It is not only through what is grown on river flood plains that export income can be earned. Today, the Nile River not only feeds Egypt and other countries that buy from Egypt, but it also provides the perfect tourist route for those who are drawn to the pyramids of ancient Egypt. The Volga in Russia plays a similar role as a tourist route through large areas of the country. Foreign tourist expenditure is the equivalent to exports in economic analysis – tourists spend money in the country (catchment) they visit, where they purchase leisure and memories.

Why catchment producers trade

The major reason that a country – or a large, well-endowed catchment – does not attempt to produce everything that its citizens want is the law of comparative advantage. This is the most basic rule underpinning (international) trade. International trade occurs because of differences in the cost of production between countries. Trade also provides greater consumer choice. The difficulty involved in understanding trade is in the difference between absolute advantage and comparative advantage. A country that can provide all commodities less expensively than a trading partner would trade if the relative costs of production for different commodities differed.

The cost of production differs because countries are differently endowed with resources, climate, the skill and size of their labour force and their stock of human-made capital. This is why China exports labour-intensive products and Australia land-intensive products.

While trade across borders is a significant component of any economy, it is normal for most exchange to occur within the economy –or, in our case, the catchment. The types of transactions we discover in our economy can be arranged in what economists call an 'input–output matrix'. This (as the name suggests) records the amount of resources (inputs) used to produce something (outputs), where the resources can come from multiple industries or sectors of the economy and the produced things go to multiple destinations (households and other sectors of the economy). By the use of matrix algebra, the data in the rows and columns of the matrix are used to calculate 'technical coefficients' (the relationship between one input and the total output of an industry or a sector of the economy). On the basis of these, the flow-on impacts (or 'multipliers') of any particular economic activity can be calculated. Flow-ons are often presented in terms of employment (and other types of) multipliers; for example, a new abattoir will result in x direct jobs in the abattoir plus y flow-on jobs as other sectors of the economy benefit from new income and additional sales (due to the initial investment in the abattoir and its operation when it is opened).

The idea is very simple to explain, but it is laborious to calculate multipliers. For example, income of a farming household is spent on both consumer goods and services (termed final demand or household consumption) and farm work requirements (inputs). To produce both sets of products, people are employed. These people, plus the businesses that employ them, earn income from the purchases such as groceries made by the farming household, income that is subsequently spent by those who provide the products to farmers. A chain reaction is underway. The flow-ons eventually dwindle away to infinitely small numbers, and that is the end of the chain.

If you have persevered this far, you would notice that our economic input–output model functions in the same way that interconnected ecological models do. Connections are everywhere. Big fish eat little fish, little fish eat even smaller fish, and they eat the smallest marine life. Big fish (and all other marine life) die, to be eaten by decomposers. These are fed upon and eventually there are new big fish eating little fish, before the former become food for decomposers. This is a simple description of a typical food web. The ecological system, if at all resilient, tends towards an equilibrium state in the short term. In the long term it is likely to be 'shocked' by outside (exogenous) natural forces (think of an ice age) as well as being influenced by the gradual processes of evolution. The so called 'balance of nature' does not mean that the diversity of genes, species, or ecosystems is stable. There is a problem in economists borrowing the concept of the balance of nature and calling it 'equilibrium' in economic markets when in fact there is not a steady-state (equilibrium) in nature. To get a little ahead of ourselves, there is not one equilibrium state in our economic world, rather there are short-term equilibriums in particular markets (say for apples in the summer of 2012) but long-term disequilibrium as new products are developed and consumer tastes change.

The economic system has some profound similarities with the ecological system, while having fundamental differences. One of the differences – and the most important – is that the time scale for change is dramatically reduced in economics. You will hear economists and business people talk of the long term while meaning 'in the next three' or 'next five' years. Ecological changes, left to themselves, take what we call evolutionary timeframes – not geological ones which are much longer – as mutation and natural selection play out. The other obvious difference is that humans, as the intelligent animal, have speech and recorded language, which has led to their enormous collective intellectual power to structure their societies and economies to their liking – something the other animals cannot do. A human economy is a human construct, with humans doing their best to obtain from nature what they want. Humans have far fewer limits imposed than those faced by the other animals – a stark case is our ability to extract and use the fossil fuels nature laid down millions of years ago.

An external force that drives major changes in the economic system can appear without warning. A technological innovation or a change in consumer preferences can come at any time; however, one would hope that not as rapidly as the yearly fashion cycle in clothing, which is a human invention that could be discarded presumably without any loss in human well-being – in fact, there could be large benefits in not feeling obliged to discard perfectly good clothing to keep up appearances. Or are the authors overlooking some innate feature of human psychology?

The relatively rapid change in economic values (whether driven by fashion or innovation) makes the task of linking the economy to the environment – as we are required to do in the pursuit of sustainability – extremely difficult. Little in nature changes quickly. There is a most obvious – and important – difference between ecological thinking and economic thinking. Economic values are not set, staid, locked-in, but intransient. They are ever-changing, often in small changes, sometimes in significant large changes. Ecologists must come to recognise this when they ask economists to 'value the environment'. The human

Box 5.2: Expect change

To give one example of how we must come to recognise the dynamic nature of human values, consider the case of Niagara Falls. In the late nineteenth century and the early decades of the twentieth century, this natural feature was considered to be the world's premier 'natural asset', in high demand for honeymoon trips. Had an economist used one of the modern techniques to measure the unpriced (non-market) value of Niagara Falls, the answer would have come out as very high. Today, with hundreds of World Heritage areas now in the public eye, its relative economic value is much diminished. It is the same falls, the same quality of water tumbling over a cliff. Presumably its ecology is basically unchanged. That has not stopped its decline in economic value.

environment in which the human economy sits is very dynamic, extremely variable, and if there is a progression to another (higher?) state (similar to Darwinian evolution), its laws are very poorly understood. One only has to look to the great poverty in the world to understand that so little is known about the process of economic growth.

As long as we keep in the front of our minds the fact that economic values change, sometimes rapidly and unexpectedly, and inevitably, we should be able to achieve our goal of making useful links between the economy and the environment. If we can't achieve this we have no hope of delivering a better, sustainable world. Recall that our task is to manage our economic decisions (what we produce and what we consume) within the bounds of the natural system. If we can't control our economic activities we are on the road to ecological disaster. The point we make in Box 5.2 we have made previously. It is highlighted here to underscore the case that our task is to use economic analysis to measure things as they stand and to be alert to the need to revalue at short notice. Of course, only if the values we calculate represent ecological reality should we use economics to make decisions. Hence the need to understand ecology so to be able to identify whether any economic change is a positive or negative result.

We can, and will, use basic input–output modelling (without resorting to the construction of a model from the ground up) of our catchment to describe it as we would a country. Having done that we will delve deeper and focus attention on the variety of ecosystem goods and services – the largely non-priced values in the catchment. This is where we need to focus to explore the interdependencies between the non-priced catchment economy and the catchment environment. In the remainder of this chapter we aim to be a little more formal, singling out water use.

Water as an example of an ecosystem good

Here we focus on one good/service which is part of the catchment economy. We go from the aggregate to the specific. We are no longer adding up the flows of goods and services of all production; rather we are deliberating on the economic characteristic of a particular good. Water is obviously valuable to all catchment residents, for personal use, for gardening, watering pet cows and horses, and much else. However, the major use in most catchments (including the one we come to focus on) is for farming – and if there is a potential conflict in water withdrawals it is between diversion of water for farming use at the expense of downstream users. This is the most obvious case where both water quality and quantity come into play. It is no accident that economic textbooks use as the 'classical' case of pollution a hypothetical factory spewing its polluted effluent into a river and a downstream fishery rapidly going broke because the fish are being killed because of pollutant.

Water, obviously as essential for farming as soils are with their nutrients and chemicals, can be, in theory at least, separated out as one of many inputs that are needed to nurture farm products. If water can be isolated from the other factors of production (the use of capital equipment, fertilisers, labour and soil quality) and the quantities varied and the results of this variation measured, we could establish in production terms the contribution by water and hence a price (value) for it. This is what economists call the 'production function' approach. When asked to estimate the value of extra or less water to farmers or any other users, this is the first tool to be considered.

In formal terms, the production function is a mathematical relationship between the quantity of something produced (the output) and the quantities of inputs required to make the product. The function in its general form is written as:

$$q = f(x_1 x_2 \ldots x_n)$$

where q = output

$x_1 x_2 x_n$ = inputs (such as labour, fertilisers, water)

f = 'a function of' (or depends upon)

We would write this in English as the quantity of output (such as wool) depends on the quantities of inputs (land, labour, water and so on). This could be set out in linear form, such as $q = a_1x_1 + a_2x_2 + \ldots\ a_nx_n$, where a_1, a_2, a_n are numbers that indicate how much output increases if the input to which they relate is increased by one unit, all other inputs held constant.

Farmers apply the reasoning behind this function when deciding on how much fertiliser, water or whatever to apply to a given parcel of land, although they hardly likely to explain it in formal economic terms and jargon. In fact, it is at the individual farm (or business) level that production functions have their most practical use. To foreshadow a frustration in our investigations in the Little Swanport catchment, we discovered vast differences between the farms in terms of levels of outputs produced and levels of inputs used. We discovered great diversity and found there was no such thing as a typical farm. When we attempted to isolate water as an input, all attempts at describing an operational, farm-wide production function failed to produce meaningful results. Only at the specific farm level was it possible to, ever so crudely, estimate the value of an extra unit of water while holding all other inputs constant (such as fertiliser, farm machinery, farm workers, and 'imported' hay or other livestock food). As each farm situation varied to a considerable extent, the idea of seeking a meaningful average relationship was not tenable. We had to seek a way around this dead-end if we were to understand and describe our catchment. We had put much faith in the idea of a catchment as a logical ecological–economic unified system, and at our first step we were facing failure.

Water in more detail

Let us put aside this initial failure and delve into water use in some detail. Water, as we know, is a fundamental requirement for life – human and otherwise. People living in a catchment, in cities or towns, have their daily water needs, plus their wants satisfied – they do not need to wash the car, but they want to, and will if the water is available. A major human activity in all river basins is farming. Water is a most basic need for farming, whether as precipitation, stored in dams, drawn from groundwater, or dispersed via irrigation. Let us concentrate on the water requirements for farming as this is less well understood than urban and industrial water requirements.

For a moment, let us return to the concept of the production function. In theory the amount of water available for a particular farm or farming in a particular catchment should (like the level of organic matter in the soil) show up in the quantity (and quality) of farm output (for example, the number of sheep that the farm can carry and therefore wool and meat produced). What economists traditionally do to obtain measures of the value of water (or fertilisers, paddock rotations, number of farm workers, or whatever farm input is being measured) is to hold all other inputs constant (for example, the hectares being used, the number of person-hours of farm work) and vary the use of water to ascertain what increases or decreases occur in output. Now, we shall put this in economic jargon. The result is the marginal value of the variable input (that is, the value of the last unit of water applied to the farm). In theory this is a commonsense, straightforward approach. In practice it is completely dependent on the availability of data that shows the relationship between water use and farm productivity – while accounting for all other variables that affect farm output. This is usually achievable for an individual farm if the farmer has records of water used during a set period, plus records of all other inputs used (such as fertilisers). Records of product sales are also required. The necessary data are usually available where farmers have to monitor their water use – and, in particular, pay for each litre used. In this situation they have a strong incentive to 'value' water and use it accordingly. Before you think why could we not rely on this type of information, we need to point out that in our case study catchment farm water use is neither monitored nor paid for. This is not unusual in farming communities. Only very recently, and then only in some countries and regions, are non-urban water users being required to pay according to the amount used. This lack of data is a major drawback for water accounting over most of the world – yet water is a precious resource that nations will fight over, and accounting for use is surely the first step in designing a system where water goes to its highest valued use.

Where water is not subject to price determination by demand and supply but rather provided free or at a subsidised price, farmers do not have the incentive to investigate the value of water – and to apply it to the most valuable crops or animals. Nevertheless, farmers do – if they have the time and financial means available – undertake rough experiments in altering the use of this or that input. Obviously, the inputs that they purchase need to be allocated carefully if profits are to be maximised; for example, purchase more fertiliser and monitor its effect on grass or crop growth; sow a larger area and measure the results; change crop varieties; and many other such experiments. This project-oriented, modern-day farming is still developing, and an older-style, more intuitive and historical approach is still common. The latter group of farmers do not experiment but rely on handed down conventional wisdom and the experience of other farmers. From knowledge passed down over generations, or from advice given by agricultural extension officers, farmers obtain an idea of what to expect if they alter the amount of water, fertiliser, or whatever. This is enough information for many farmers. Experiments involving changing farm inputs can be both costly and risky. The vagaries of nature and markets provide enough risk for most farmers.

Economists cannot go to farmers and ask them to put on hold their tried and true use of inputs such as water or fertilisers while an experiment is undertaken. Hence, a reliable, non-experimental method to value the benefits of the application of more (or less) of a farm input is required. The necessary data to start the valuation process exists in farmers' records. Farmers are approached and asked to open up their financial books to the researchers, who hope the farmers have kept reliable and comprehensive records (if for no other reason than the taxation authorities require verifiable data on costs and earnings).

For the economists to model a farmer's behaviour, the economist will need to obtain financial data from a large number of farms, and usually covering a number of years of farming. This means that considerable amounts of economic data need to be gathered from each farmer and, ideally, hundreds of farmers would be involved in the research. This is a statistical requirement. Small numbers are prone to low accuracy. It usually takes some effort by the researcher to convince a farmer of the utility of providing confidential financial data. Farmers (and others) tend to be wary of providing financial information (except, presumably, to their accountant), and they hesitate to use valuable time to answer extensive questionnaires. The amount of the data required for a typical study of the kind being discussed can be gauged from the lengthy questionnaire we used in the Little Swanport research. A farmer is likely to spend some hours completing the survey, the value of which he or she doubts. In our field work we visited the catchment many times; first to meet with a selection of people who we had ascertained were active in the community; the second time to interview farmers and other landholders who had not responded to a mail-out questionnaire. At this stage we engaged in face-to-face interviews or distributed a replacement questionnaire to those who were not home; this process was repeated some months later in an attempt to obtain a high response rate. Our overall response wate was 65 per cent; however there was a bias as we received just under a 50 per cent response from the small number of very large farms. As explained later, the increased severity of the drought in 2007 led us to revisit a group of randomly selected farms to ascertain the changed economic circumstances.

There is an expectation with the type of research being discussed here that a production function representative of all farms in an area (such as a river basin) can be derived from the data provided by the farmers interviewed. As explained above, a production function shows the inputs required to produce a given output – how much land, water, fertiliser and so on to produce a tonne of wheat or a bale of wool. To derive a representative production function requires that, from a mass of data, the exact contribution of the input in question (water in our case) can be identified and measured in isolation from the other inputs. To explain this in its simplest form, assume two farms are exactly the same in all regards (size, farmhands employed, crops grown, animals grazed), and one uses more water than the other. We would expect the farm using more water to produce more farm products, and hence earn more income.[1] The economic value of water would be staring us in the face. So much for theory, and so much for a taste of what we wished to attempt in our study of the Little Swanport catchment. In the next section of the book we will broaden, considerably, the subject matter and return later to how we solved the valuation problem.

6

The world's great river basins

Amy White and Tor Hundloe

Imagine a river basin

All of us live, work, and manage our affairs in what we call a nation or country. We expect to find rivers and river basins, if only the most meagre ones, in all countries. The tiniest amount of rain needs to drain somewhere – unless the soil or sand is so porous that it disappears without trace. Draining water forms a rivulet. A number of rivulets fed by the same rainfall combine to form a proto-catchment or rivulet basin. Australia is a country containing many significant river basins, one large and dominant, plus hundreds of smaller ones. Our challenge, and your challenge, is to imagine life in a river basin as life in an ecologically defined nation. Imagine you live in a catchment. You do your business in a catchment. Your government is formed by voters in the catchment. You trade with people in other catchments. Of course this is all hypothetical; yet imagine how much easier environmental management would be. No intra-catchment negotiations would be required.

We are going to select a few catchments across the globe with the objective of highlighting their economic importance (as political entities) rather than their ecological importance, which is the point of interest we have just lauded. Of course, we do not disregard the latter – in fact, it is paramount – but we want to emphasise another way of understanding catchments, one which in an ideal world is also likely to have merit.

There is a seemingly infinite and ever-changing number of rivers scattered throughout the diversity of the world's landscapes, each with its own unique story to tell. In reality there is a finite number, and the differences are relatively easy to map with modern satellite technology. However, one solitary human casting his or her eye over the totality of the globe is easily overwhelmed by the number of rivers, rivulets and channels that seem to change size, colour and direction (although always downwards) according to each rain storm. Many of these rivers and their associated catchments have played an integral role in shaping the historical, cultural and economic heritage of villages, cities and countries, over the entire globe.

For most of human history, nature has had its way regardless of what humans have done. However, as our numbers have increased and our technologies advanced, we have played a distinct and often dramatic role in shaping the natural environment and the ecosystem dynamics within the world's catchments. In this chapter, we describe (ever so briefly) some of the world's great catchments and focus on the unique human–environment interactions that have taken place throughout time in each. The treatment is, we hope, sufficient to create interest in thinking about river basins as the logical geographical unit for the human–environment interface. We recognise that in formal ecology it is conventional to divide the natural world into a number of ecosystems, and that is certainly appropriate for the task that textbook authors have in mind. Our perspective is different, in that we suggest that what links mountains, grasslands, wetlands and estuaries is the flow of water, which first meets land in a catchment.

The Nile River

Having its source in the small country of Burundi in Central Africa, and stretching 6695 km north to reach the Mediterranean Sea in the 250 km-wide Egyptian delta, the Nile is the world's longest river. Hence, its pride of place! Two main tributaries join the stream, the White Nile, beginning in equatorial East Africa, and the Blue Nile, which flows from its source in the Ethiopian highlands to join with the White Nile at the Sudanese capital of Khartoum. The White Nile rises at Lake Victoria in Tanzania and flows north through Uganda, Sudan and eventually Egypt. North of Cairo, the Nile splits into two branches that feed into the Mediterranean Sea, forming the large Nile Delta. The catchment area covers approximately 3.1 million km^2, including 81 500 km^2 of lakes and 70 000 km^2 of swamps. This is a big river, with a big history.

There are eleven riparian countries within the Nile catchment: Burundi, the Democratic Republic of Congo, Egypt, Eritrea, Ethiopia, Kenya, Rwanda, South Sudan, Sudan, Tanzania and Uganda. These countries were carved out of Africa by colonial powers, not just recent ones such as the Belgians, Italians, French, English and Portuguese, but by ancient Phoenicians, Persians, Greeks, Romans and, in medieval times, Islamic colonial powers.

The Nile is an African river. African-sized populations are involved. The catchment is home to a population of around 160 million, most of whom reside in the 25 large cities within the catchment.[1] The population is growing rapidly as this is a part of the world where large families are normal and there is an economic imperative to produce children as workers. Not all are expected to survive to adulthood. One of the tragedies of much of the African continent is the need to have large families due to the high rate of infant mortality and the fact that subsistence agriculture needs many hands. Thomas Malthus got his predictions of the human population outstripping agriculture production right, if only for Africa. We expect he will be proven correct for the planet as a whole in the near future, as the limits of arable land and fresh waters are reached.

The landscape along the Nile varies from rainforests and mountains in the south to savannas and swamps in South Sudan, and finally huge deserts throughout Egypt. If you ever wonder about the bareness of the Egyptian deserts (called as-Sahra-al-Gharbiyah in the west, as-Sahra-al-Sharqiyah in the east), reflect on the fact that 99 per cent of the Egyptian population inhabit the Nile valley and delta, and from the tiny area of land forming the river flats of the Nile all Egyptian food is grown. It is as if a tiny strip of land on either side of Murray River had to provide food for all Victorians, South Australians and a sizeable population of New South Wales people. The lower Nile river basin is in reality Egypt. Only in the wide delta does the country become something other than a thin stretch of water and narrow flood plains.

The first major civilisation along the banks of the Nile developed around 5000 BC in the area now known as Egypt. This predates by millennia the development of Greek science and philosophy (circa 500 BC). Most of what we pay attention to today, such as Archimedes' principle, the Hippocratic Oath and the philosophy of Socrates, dates from this more recent period. Egyptian civilisation is really old in our scheme of things. The annual flooding of the Nile River spread fertile, nutrient-rich mud over the flood plain, making agriculture possible. The early Egyptians developed the skills to control seasonal flooding and installed primitive, but effective, irrigation. Irrigation has been crucial wherever agriculture has been practised – except where there is high rain or snow fall. The Egyptians built dykes and terraces to grow crops such as wheat, emmer, barley, flax, onions, lettuce and pulses. As Egyptian agriculture developed and a surplus was generated, the ruling class found itself with enough food to maintain an enormous slave labour force, sufficient to build the pyramids. The Egyptian Old Kingdom, which includes the first Pharaonic dynasties, had the Stepped Pyramid of Saqqara built, considered to be the oldest building in the world. Later the pyramids of Giza were built. By the time of the Middle Kingdom (circa 2040 BC) the Egyptians traded with Crete (now part of modern Greece), Palestine, Syria, Babylonia (now Iraq), and Anatolia (now Turkey). Here we have the essential ingredients of a modern economy: the division of labour (in this case between a ruling class and salves), the generation of a surplus to be used for, among other things, public buildings and trade. None of this would have been possible without the existence of the river Nile – the Egyptians would have been desert dwellers roaming the great expanse of North Africa.

Ancient Egyptian civilisation fascinates modern people. There would not be an educated person who has not read of the pyramids and the Sphinx. Many have found their inquisitiveness so strong that they have made a visit to Egypt and travelled down the Nile. Cleopatra is better known than Boadicea – if only because Elizabeth Taylor played Cleopatra in an epic Hollywood drama. More famous than Cleopatra is Alexander the Great; his mark on Egypt exists today with a city named after him. He conquered the country in the fourth century BC.[2] It had been a Persian province from the sixth century BC. After Alexander's death, the Ptolemaic pharaohs ruled until 30 BC. In 672 Egypt was captured by the Arabs. Space does not permit it, but it is possible to describe ancient Egypt and its economic and social life in more detail than we could describe twenty-first century North Korea, cut-off as it is from the rest of the world.

Let us come to modern Egypt. A major transformation of the lower Nile began with the construction of the first Aswan dam in 1899. This was 30 years after the Suez Canal was built. The dam allowed the Egyptians to control the flooding of the Nile valley and increase the amount of available farming land. The economy, still a subsistence one, grew. However, nothing of dramatic significance happened until a turn of events in the middle of the last century. Grand ideas were the order of the day when Egypt separated from Britain and was proclaimed a republic on 18 June 1953. The end of colonisation was associated with understandable optimism, leading, on occasion, to risky adventures. Egypt's second president, Gamal Abdel Nasser, nationalised the Suez Canal Company in 1956 in order to obtain the funds to construct the proposed new Aswan Dam. Both projects were very risky. Had the United States not viewed Egypt as crucial in its ideological war with the Soviet Union, the nationalisation of the canal might have been overturned by military action from Britain and France. The Aswan Dam was a risk of another kind: what were the environmental and economic consequences of damming the world's largest river?

With the assistance of the Soviet Union in 1960, the Aswan High Dam was constructed further upriver than the old dam. Today, the Aswan High Dam impounds one of the world's largest reservoirs, Lake Nasser, and generates over 10 TW of electricity per year. This is an enormous success – and it happened within the borders of 'the last country in line' – that is, at the end of the river, thereby not impacting on downstream countries. The question 'what if' is tantalising. What if one or a number of the upstream countries had built large water storages? What type of country, economy and society would Egypt be today if starved of water?

In the era when the Aswan High Dam was constructed there was little serious thought about the environmental and social consequences of such a significant interruption in natural water flow. Ancient sites such as Nubia were submerged, and large-scale resettlement of people was necessary. That was decades ago, and it is now only a footnote in modern Egypt's history. There have been considerable unforeseen environmental and social problems. Today, large stretches of the river are clogged with silt, and the resulting decrease in natural flooding has led to a reduction in the productivity of downstream farms. Large areas have been damaged by erosion and salinity.

Was the Nile dam worth it? The Nile irrigation system supports agricultural production on over 5.5 million ha of land within the Nile valley. It is noteworthy that less than 3 per cent of Egypt is arable land, and that only 0.5 per cent is under permanent cultivation. This is a tiny amount of land to feed a large population. The regulation of the flow that was made possible with the construction of the dam increased agricultural production enormously in this small area. This is what dams and irrigation systems do in the short term. We are still pondering their long-term effects.

The major products of the Nile flood plain today are wheat, rice, sugar cane, cotton, vegetables, melons, sorghum, citrus fruits and various legumes. This is a great variety of foodstuffs, mostly for domestic consumption. It is relatively inexpensive to produce. Water is virtually free – the Egyptian government has to maintain the dam – and labour is still relatively inexpensive. The agricultural production on the flood plain accounts for approximately one-third of Egypt's national income of about $300 billion annually. The river also provides a range of seafood, including Nile perch and tilapia.

Egypt is the Nile River – no other country is as one-river as Egypt is. It was in Cleopatra's day; it is today.

The Amazon River

The Amazon River is fed by hundreds of smaller rivers, draining an area of more than seven million km^2 (which is more than twice the size of the Nile catchment), and discharges around 30 trillion litres of water per day. More than two-thirds of the Earth's fresh water is in the Amazon basin. The Amazon is the world's largest river in terms of catchment area, number of tributaries (over 1000), and the volume of water discharged. Only in length is it second to the Nile. The Amazon stretches 6500 km from the eastern edges of the Peruvian Andes to its wide estuary at the Atlantic Ocean. Much of the river passes through the dense Amazon rainforest. The Amazon rainforest covers approximately one-third of the South American continent, yet another very important fact. Rainforests are the prime source of terrestrial biodiversity, and they are diminishing in size daily; this because the ecosystem goods and services we obtain from them are, in the most part, not valued in economic terms. Our urgent need to value the unpriced is most obvious with rainforests. We have a better estimate of the number of stars in the galaxy than of the number of species in the Amazon. Very few have been identified. Concerned, as we must be, with the planet's shrinking biological diversity, the Amazon rainforests are an increasingly valuable resource. The law of supply and demand suggests nothing less; however, this does nothing to save the next hectare from the chainsaws and bulldozers.

More than half of the Amazon River runs through Brazil. The other countries it transgresses are Peru, Ecuador, Bolivia, Venezuela and Columbia, more than enough for there to be conflict over the allocation of water if it came to that. So far this has not been a significant issue. A major reason for this fortunate situation is that the catchment is not heavily populated, with only 26 million people living in it, and the majority of these live at the mouth of the river. Demand for water upstream is not what it would be if the river had to provide for large populations in a number of cities. Fortunately, there are very few cities scattered throughout the catchment, with the cities of Manaus and Belém the largest.

The dense rainforest vegetation that covers much of the Amazon basin has made it near impossible for traditional forms of exploration, and many of the tributaries of the Amazon River remain untravelled and untouched by modern people. In certain remote areas, the few remaining Indian people (estimated at below one-quarter of a million) continue to live subsistence lifestyles, engaging in primitive agriculture, hunting, gathering and fishing. The ecosystem goods and services provide the basics of their livelihoods.

Today the Amazon basin economy, in the areas where the land has been cleared, is dominated by soy plantations and cattle ranching. Soy and maize have come to form the basic ingredients of manufactured food, and as it increases in demand so is more land converted to growing these crops. Timber harvesting (Brazil has the third largest timber resources in the world) is the major extractive industry, and the conventional wisdom is that rapid, unsustainable clearing is the greatest threat to the basin and the vast diversity of non-human animal life that depends on it. As the road network expands through the rainforests, there are increasing concerns about the flow-on effects induced by human activities – as the forest canopy gives way to roads, timber plots and farms, the soils becomes susceptible to erosion and productivity declines. The future is uncertain and a genuine concern of global significance. We can be fully optimistic that Brazil is poised to become an economic power of considerable magnitude, but to what extent will entrepreneurs destroy its most important asset in the process?

Is there an economic saviour for the Amazon? River trips and rainforest tourism cater for a small but growing adventure market. There is hope – but so far not sufficient action – that ecotourism could save the forests. Ecotourism provides a steady flow of income and jobs compared to unsustainable timber-getting. Done well – with numbers of visitors limited and prices set high so to maximise income for the country – ecotourism has great potential. Australia has proven the income-generating ability of ecotourism in a rainforest setting, the Wet Tropics (aka Daintree) World Heritage Area surrounding the tropical city of Cairns.

The Amazon River has an iconic status. It is home to a huge diversity of flora and fauna; for example there are more than 3000 species of fish in the river. One is known universally for its eating habits. What child has not been intrigued by the human-flesh-eating piranha? Intrigue with the unusual could be used as a draw card for tourists.

Farming of wild rubber latex, cacao and Brazil nuts provide the principal exports from the catchment, with exotic forest products (resins, barks, feathers) of minor importance. If demand became strong for the ecosystem goods and services in the Amazon jungle, it would be saved. Pharmaceuticals are high-value end-products. Bio-prospecting, if it delivered results, could be an ongoing source of royalties.

Extractive industries could continue as long as the scale did not expand. These are the possibilities. The reality is that, today, not only forestry threatens the natural environment; clearing of land for soy bean plantations and ranching continue to convert jungles to monocultures. The conventional view is that every year an area the size of Belgium is cut out of the Amazon rainforest for agriculture. This is not sustainable. The campaign to protect the Amazon forests from land clearing was made famous by the protests of an indigenous rubber tapper and environmental activist, Chico Mendez. In an attempt to preserve the Amazon and the traditions of his people, he organised fellow workers into the National Council of Rubber Tappers. They went on to protest against the land clearing. The international media took interest and 'celebrities' who like to be associated with good causes helped sustain the interest. The protestors' efforts were partially rewarded with the government's support for the creation of 'reserves': small parcels of land put aside for sustainable land uses. Then, in 1988, Chico Mendez was murdered. His assassination by those with financial interests in exploiting nature only served to focus greater attention on the cause for which he fought. He became a martyr. Today, his co-workers and activists continue to uphold his legacy. There are now more that 20 reserves within the Amazon basin, covering over 32 000 km^2, and providing refuge to over 3000 families of indigenous tappers and farmers. However, note that there could have been 10 million indigenous people living in the basin before the Europeans arrived. The battle is far from over. The Amazon is a mighty river with a massive catchment. The fact that the population living in or using the catchment is relatively small will help to preserve it. The eyes of the world are on the Amazon.

The Yangtze (Chang Jiang) River

Beginning in the snow-covered Tibetan Plateau in the west, and flowing 6300 km to the mouth at the East China Sea, the Yangtze (or Chang Jiang) is the longest river in Asia and the third largest in the world. The river is the natural divide between north and south China. Shanghai, China's most populated city, is the gateway to the Yangtze, and is the country's major port and commercial centre. Modern Shanghai is one of the great cities of the world – a monument to China's towering economic success in the twenty-first century.

The Yangtze catchment is of tremendous economic importance to China's agricultural and industrial sectors, and in years to come will play an important role in tourism. The rich alluvial soils and the high rainfall in the middle and lower catchment make this area the granary of China, a country of 1.3 billion people. More than one-third of the population rely in some form or another on the Yangtze. In population terms, the Yangtze river basin would be the world's largest economy (or 'country') if we drew a border around it. Of the 1.8 million km^2 of land within the catchment, around 47 per cent is cropland, an area that produces over half of the country's rice, as well as cotton, tea and oilseeds.

The numerous dams along the river provide irrigation for surrounding farms and help to control unwanted flooding. Add to this the fact that the Yangtze, plus some of its 700-odd tributaries, form China's main navigable waterway. The river serves as a transportation network for cargo, residents and, increasingly, tourists. Large ocean vessels can reach as far as 1000 km upstream to the industrial city of Wuhan, with smaller vessels making it 1600 km to Yichang. The recently constructed Three Gorges Dam, west of Yichang, will allow freighters to navigate as far as Chongqing, 2250 km from the mouth of the river. This is, by European standards, an extremely long 'freeway' over which to transport produce and people. Unlike human-made roads it requires virtually no maintenance. Until the later part of the twentieth century, naked men struggled along the river's shore hauling barges and not just along the calm stretches but also up the rapids.

The Three Gorges Dam is four times larger than the Hoover Dam near Las Vegas, the Hoover being the world's benchmark. The Three Gorges Dam incorporates the largest hydropower plant in the world, with a total installed capacity of 18 200 MW. Its reservoir when full will be 600 km long, and the water level will be raised 175 m. Being a multipurpose dam, it has the capacity to reduce the impacts of flooding on surrounding farms and urban areas in the middle and lower catchment. These are undoubtedly significant benefits. However, despite these and the other positive impacts of the dam, there is considerable concern about the long-term environmental impacts on surrounding areas. It is not the immediate impacts that are of most concern, but rather the induced ones (the expansion of industries and cities and their consequent pollution) that are raising second thoughts! Already there has been severe dislocation of people with unmeasured social impact. As a result of the construction of the dam, 40 000 ha of fertile farming land have been lost, and in the order of one to two million people forced to move from their homes. These are the poor of China, those forgotten so far in its rapid growth.

No simple cost–benefit analysis could accurately capture the total costs and benefits of this dam. And no cost–benefit analysis, simple or otherwise, is possible, given the lack of reliable data. The type of detailed economic data gathered in the rich countries (by their economic and environmental agencies) is not gathered – or if gathered, is not reliable – in China. In rapidly developing countries such as China, the degree of optimism and self-belief is so elevated that benefits are readily assumed and externalities are not considered.

There are nine large cities within the Yangtze catchment and one-third of China's population live in the catchment, mainly in the river valleys and along river banks. Farming, logging and high density urban development have already resulted in serious environmental degradation. Land clearing and erosion have greatly increased the sediment load, impacting on water quality and making certain areas more prone to flooding. These problems are likely to accelerate.

Should the dam have been built? Of major environmental concern today is the estimated 23.4 billion t of sewage and industrial wastes dumped in the river every year. The Chinese government is attempting to clean up the river. We have no idea if it will be successful. However, with appropriate and existing technologies this is clearly possible. If the clean-up is serious and the best technology is used, the river could be returned to something close to its natural state as the Thames and numerous other European rivers were some decades ago. It is simply a matter of political will, including the willingness to spend the money required.

If we were to attempt to model the ecological–economic–social dimensions of the Three Gorges Dam with the aim of coming to a judgment of its worth, it would be as if we were modelling the whole Chinese economy. This is because of the vastness of the catchment, the population residing and making a living within it, and its economic flow-on to the rest of the Chinese economy and nation. As much as we would like to do this, modelling the Chinese economy, in whole or part, is beyond the expertise of the best economists, primarily because of this economy's dynamic nature and the still poor-quality statistics on some sectors.

The Mississippi River

The Mississippi River would appear to commence as a small stream draining Lake Itasca in Minnesota, and flow 3780 km south to its mouth in the Gulf of Mexico, Louisiana. However, a closer inspection of the river's source suggests a beginning in the neighbouring country of Canada. Water drains from Canada into the headwaters of the river, and therefore its catchment extends into that country. Together with its major tributaries (the Missouri, Ohio, Illinois and Arkansas rivers), the Mississippi forms the largest catchment in North America. The majority of the United States could have been called the United States of Mississippi, so large is the catchment. Over 40 per cent of the Mississippi's water comes from the Missouri River (with its source in the Rocky Mountains). It joins the Mississippi at St Louis, bringing with it large volumes of water and sediment. The Ohio River joins the Mississippi south of St Louis, west of Nashville. The Arkansas River (with its source near Denver) joins it south of Memphis, and the Red River joins relatively close to New Orleans. The Mississippi catchment drains an area of 3.2 million km^2, which includes 31 American states and two Canadian provinces.

Native Americans made use of the river throughout their period of settlement of the continent. The flood plains were farmed and the river fished, and it was a major transportation route. When the Europeans came it became central to the exploration and economic development of the nation. The Mississippi and Missouri provided ready-made freeways for the early European settlers as people and goods literally flowed throughout a large part of the United States on the river system. The paddle-steamers, the cardsharps and the riverboat girls were the equivalents of the wagon train, the cowboys, the outlaws and the bar-room girls. Mark Twain describes the heroics of Tom Sawyer and Huckleberry Finn in some of the most delightful American literature, with these two characters (and their adventures) introducing the Mississippi to generations of school children.

The river has been dredged over a very long period, and shipping channels constructed to allow large vessels to move freely, making the river a key route for bulk freight. Petroleum, iron, steel, grain, rubber, paper and wool are among the major products shipped at the lower end of the river. Major ports along the river include those in New Orleans, St Louis and St Paul. Today, the Mississippi basin is of immense economic importance to America, producing 92 per cent of the nation's agricultural exports and 78 per cent of the world's exports in feed grains and soybeans. The rich alluvial plains of the Mississippi River provide ideal growing conditions for soybeans and feed-grain farms in the north, and cotton, groundnuts (peanuts) and rice in the south. There is also an abundance of fish in the river, which are harvested by both recreational and commercial fishers. In an input–output transaction table of the US economy, the Mississippi basin would dominate the agricultural sector with its forwards and backwards links to other sectors of the economy.

High-intensity agriculture, urban development and manufacturing industry have had a major impact on the water quality of the river, as it is used to dispose of municipal and industrial wastes. The human impacts in the catchment suggest that its days of pristine environmental conditions are gone forever. However, it will undoubtedly remain one of a handful of productively powerful river basins of the world.

The Yenisei River

Out of the 10 longest rivers in the world, Russia has three; it has four out of the 11 longest. Russia has rivers flowing into the Arctic Ocean, the Pacific Ocean, the Caspian Sea (the Volga and Ural rivers), and into enormous inland lakes. The Yenisei is one of these rivers. Formed by the union of two headstreams in the Sayan Mountains in Russia near the Mongolian border, the river flows 5540 km north-west through the great expanse of central Siberia. Emptying into the Yenisei Gulf, an arm of the Kara Sea, it is the largest river system flowing into the Arctic Ocean. Its catchment contains some 20 000 tributaries, including the Angara, Selenga and Ider rivers, which collectively drain an area of approximately 2.6 million km^2. This is an enormous area – more than two-thirds of the Mississippi drainage basin – yet very few people outside Russia know of its existence.

A large range of climatic and physical conditions are found in the catchment, from sub-Arctic in the north to continental conditions in the middle and southern parts. Consequently the river supports a wide range of economic activities and lifestyles. In the north, a relatively small population live traditional sub-Arctic lifestyles, and their economy is based on hunting, fishing, breeding of reindeer and fur farming. There is a small amount of nickel and copper mining in the catchment, with abundant mineral deposits found along the banks of the middle section of the river. There is an established logging industry. Only in the southern section is there extensive industrial development. For example, power stations have been built at various locations along the Angara tributary, with a combined generating capacity of 25 GW. Where there is intensive forestry, industrial and mining industries there are dramatic impacts on the water quality of the Yenisei and its tributaries, with the Angara River being the most heavily polluted.

There are only five large cities in this enormous catchment and much of the area is relatively unpopulated. The major cities include the port of Krasnoyarsk and the industrial city of Irkutsk. As with many isolated parts of Russia, we have too little first-hand or empirical knowledge to assess the environmental damage and (let us be even-handed) the economic benefits that are generated in the catchment. The massive size of Russia and the natural resources it contains will mean that in the future we will come to know much more about the very large, eastern river basins in the country. While all attention is on China and India where labour is cheap, Russia, with a relatively small population, good technical education base and fabulous natural resources, is capable of becoming a genuine economic super-power.

The Huang He

The Huang He (Yellow River) is the second longest river in China, stretching 5464 km from the high plateau of Bavankela Mountain in Qinghai Province, to its mouth at the Bohai Sea in Shandong Province. The catchment area covers around 750 000 km^2, with a mean annual runoff of 58 billion m^3, 90 per cent of which is used for agriculture. The river carries the highest sediment load of any river in the world, comprised mainly of fine-grained calcareous silts eroded from the upper regions. The load is coloured yellow; hence the name, which translates to 'Yellow River'.

The Huang He played a key role in the early development of China, and is considered the cradle of Chinese civilisation. It is the Euphrates–Tigris of China, the Nile of China, the Indus of China. Villages formed that ultimately became cities. Today the river runs through some of the oldest cities in China, including Xi'an and Lanzhou. The catchment area covers nine provinces and has a population of almost 200 million people, nearly half of whom live in flood-prone areas. This is in the order of one-sixth of the Chinese population. One cannot underestimate the economic and social importance of the Huang He and its catchment.

The soil on the river plain is made fertile by the river's muddy silt. It is ideal for farming, and the flood plain is a key resource of the Chinese agricultural sector. The major products grown are rice, millet and wheat. However, as a result of the rapid industrialisation and urbanisation of China, large areas of the rich soils are becoming unsuitable for agriculture. There is severe water pollution, changes in water flows that are causing parts of the river to dry up, and general loss or degradation of land through urbanisation.

Silt deposition over millennia and the more recently constructed dykes have caused serious floods in recent times. Major floods have caused some of the highest death tolls from natural disasters in history; so devastating have they been that the river has been called 'China's sorrow'. The worst flood in modern times occurred in 1931, when some 88 000 km^2 were completely flooded and 80 million people were made homeless. The flood itself and the famines and epidemics that followed caused the deaths of around one million people. For all the sorrow the river can bring, if it did not exist China as we know it would also not exist – most importantly there would not be enough food grown to feed one billion plus people.

The Ob River

The Ob River is another of Russia's little-known rivers. It twists its way across western Siberia from its origins in the Altai Mountains, through the Gulf of Ob and into the Kara

Sea of the Arctic Ocean. It flows 3700 km north-west, passing through a diversity of landscapes including steppe, taiga, swamps, tundra, and semi-arid landforms. The catchment consists of over 1900 rivers and drains an area of around 3 million km^2, only marginally less than that of the Mississippi catchment. The Irtysh River is its major tributary, bringing water from parts of China and Kazakhstan.

The catchment has a very small population of around 27 million people, the majority of whom reside in the 17 large cities within the catchment. Novosibirsk is the major centre, and the third-largest city in Russia. The Ob is principally used as a transportation route for cargo, mainly lumber and grain. Ice during the five to six months of winter can hinder the passage of vessels. Together with the trans-Siberian railway, the river allows the movement of cargo in and out of the Ob catchment. Novosibirsk, Narnaul, Kamen-na-Ob and Mogochin are the major ports along the river. In comparison to the relatively minimal industrial activity on the other major river systems in Siberia (the Lena and Yenisei), industry is well developed in the Ob catchment cities such as Omsk and Barnaul, which are major manufacturing centres.

The catchment's overwhelming economic significance in the present era is based on fossil fuels. Two-thirds of Russia's crude oil and natural gas output is extracted from the taiga and tundra zones of the middle and lower Ob. This underpins the great importance of this river – oil and natural gas have become the backbone of a resurgent Russian economy after the virtual economic collapse following the turmoil of the Yeltsin coup that effectively removed Gorbachev and the Russian parliament in 1991.

During spring, the steppe zones in the southern catchment are covered with wheat. Major hydro-electric plants are located on the Ob itself and the Irtysh tributary, with a combined capacity of an immense 250 TW. This is 25 times the amount generated from the Aswan Dam. While these facts point to the economic significance of the Ob, there is a downside. Severe pollution from surrounding land uses has had a detrimental effect on the formerly famous fisheries in the lower Ob.

The Paraná River

Formed by the union of the Paranáiba and the Rio Grande rivers in the plateau of south-central Brazil, the Paraná flows some 4880 km to form the second-longest river in South America. The river merges with the Paraguay River near its mouth, forming the Rio de la Plata estuary at the Atlantic Ocean near the Argentinean capital of Buenos Aires. The catchment covers an area of approximately 2.8 million km^2 that encompasses most of south-eastern Brazil, Paraguay, eastern Bolivia and northern Argentina. In the upper catchment, the river flows in a deep bed in steep country, resulting in many waterfalls. The lower catchment is characterised by shifting channels and flooding.

The river is the principal transportation route for the interior south-east of the South American continent. Dredging has made it possible for large ocean-going vessels to reach as far as Rosario, Santa Fe in Argentina, and smaller river craft as far as the Iguaçu River. The main cargoes are petroleum, agricultural produce and manufactured goods. In 1982, in the stretch of river between Paraguay and Brazil, the Itaipú Dam was constructed for hydro-electricity production. This resulted in the submersion of the picturesque Guaíra Falls. A large reservoir was formed and significant environmental and social effects on the surrounding area occurred.

The majority of people live within the 19 large cities located along the banks of the river, its tributaries, and along the Atlantic Ocean coastline at its mouth. These include Buenos Aires, the capital city of Argentina, and the heavily populated city of São Paulo in

Brazil. Although large areas of the catchment remain in pristine, natural condition, the pollution of the highly populated areas are a major risk to the environmental health of the river and the downstream ecosystems. The river is home to a diverse range of fish which provides local people with an abundant source of protein. Although this resource is yet to be over-exploited, population growth and increased demand are likely to result in over-fishing.

The Congo (Zaire) River

Rising in the highlands of north-eastern Zambia, the Congo (Zaire) River flows 4700 km to reach the Atlantic Ocean at Banana in the Democratic Republic of the Congo (DRC). With a catchment area of approximately 4.1 million km^2 (well over the size of the Mississippi and much more than half of the Amazon), it takes in most of the DRC, the Central African Republic, eastern Zambia, northern Angola, and parts of Cameroon and Tanzania. It is located in the heaviest rain-belt of Africa and is second in the world only to the Amazon in terms of the volume of water discharged. It truly is a mighty river. The equatorial climate that prevails across a significant part of the catchment supports a dense evergreen forest of extremely high biodiversity value. Unlike most rivers, there is no delta at the mouth of the Congo.

The river is made up of a dense, branching network of tributaries, sub-tributaries and small rivers, which together form the most important navigational system in Africa. Dredging is required to allow ocean-going vessels to reach the major sea port of Matadi (which is 148 km from the coast), at which point trains are used to bypass the rapids of the lower Congo and allow cargo to move between Matadi and the major industrial hub of Kinshasa. Smaller vessels are then used to transport goods as far as Ubundi in the upper catchment.

Subsistence agriculture remains the main source of food and income for most people living in the catchment. This is sub-Saharan Africa, where notwithstanding very significant economically valuable resources, both conventional economic growth and human development remain more elusive today than when the colonial powers departed 40 years ago. Most of the countries of sub-Saharan Africa suffer from poor government, when not corrupt government, a lack of social capital, and ethnic conflicts of the most horrendous kind. National boundaries were arbitrarily drawn by the colonial powers with no regard to tribal affinity, and to this day tribal loyalty remains supreme. In this part of the world we cannot assume that national boundaries drawn on a river basin basis – a model we believe would have aligned political power with ecological reality – would have lessened the ethnic problem.

Subsistence or local-market produce rules except where the global giant corporations have taken hold of land – or of farmers' produce. The crops grown for local consumption include the staples such as cassava, corn, yams, rice and pulses. Rubber, coffee, tea, sugarcane and palm products are grown commercially, mainly for the export market. Rainforest timbers are exported, and this is degrading the still near-pristine natural environment. The river provides an abundance of fish, most of which is consumed locally, and the adjacent plains support grazing of goats, sheep and cattle, providing meat and milk for local tribes.

Significant areas of the catchment are being exploited by the mining of copper, cobalt, zinc, manganese, gold and silver. The mining industry, powered by numerous hydro-electricity facilities located along the banks of the Congo and its tributaries, is a growing environmental concern. The hydro-electric potential of the river is far from being fully exploited. Plans are on the drawing board for major expansion as the threat of climate change has brought hydro-electric power back into serious consideration as a clean source of energy to fuel the future economic development of sub-Saharan Africa. This is notwithstanding the inundation of large areas through the construction of new dams. Maybe this price will be paid for clean energy.

Of all the world's regions, sub-Saharan Africa has the largest population growth – a predicted four-fold increase in the next 40 years. This also applies to the Congo, where population is growing at a rapid rate, approaching 3 per cent per annum. The abundance of natural resources in the catchment precludes a Malthusian collapse – population growth outstripping food production – for a considerable time. Yet, despite its natural resources, much of the population lives in poverty due to political disharmony and armed conflict, an over-reliance on fluctuating export markets for crops, and a lack of foreign investment. The latter is too risky. Recently, more than 200 million ha of logging titles were issued to international timber companies, with the promise that this would assist in the economic development of the region. However, timber-getting is creating environmental degradation and social discord in the Congo basin. The economic benefits end up in the hands of anyone except the local poor farmers or their (often poorer) cousins in the city slums.

There are, as anyone who watches television knows, a litany of sad stories in sub-Saharan Africa. The fate of the Congo River and its people does not have to be yet another; however, the signs are not good. The promise of the United Nation's Millennium Development Goals, the exhortations of debt-forgiveness by people like Bono and development aid are not going to be enough. Tribal and religious warfare is waiting for the next excuse – which can be as innocuous as an accidental death – to be rekindled from the smouldering embers of the last bloody conflict. Our description of the Congo catchment does not reach beyond this unhappy prospect for the fear of raising expectations that cannot be fulfilled.

The Lena River

Rising on the western shores of Lake Baikal in Siberia, the Lena River flows some 4400 km to its mouth at the Laptev Sea, where it forms the largest delta in the Arctic. The river is fed by over 2500 tributaries and covers an area of around 2.6 million km^2, which is comparable to the Yenisei catchment on the west of Lake Baikal, and over three-quarters of the size of the Mississippi catchment. The Lena is another great Russia river, destined to become more important in the future.

Biodiversity defines the Lena. Its delta is a critical breeding ground for numerous migratory birds, as well as a diverse range of fish. Spanning some 61 000 km^2, it is the most extensive protected wilderness area in Russia. The barren landscape of the Lena catchment is very sparsely populated, with a density of less than one person per km^2. This is a good reason to believe that for the foreseeable future human population will have an insignificant adverse impact on the Lena wilderness.

Yakutsk is the only large city within the catchment and it serves as the major port and business district. With average winter temperatures of –40°C, and lows of –64°C, it is one of the world's coldest cities. Compared to other cold places such as the Norwegian city of Tromsø, which has an average winter temperature of –12°C, and lows of –18°C, one can only attempt to imagine the stark, bleak and extremely difficult environment faced by inhabitants of Yakutsk.

The river is a transportation route for cargo ships during the summer months and is navigable for some 3380 km. Timber, furs, gold, a range of industrial goods and food are the main products transported. During the warmer months, the flood plains of the lower catchment are used for growing crops such as oats, barley, wheat, potatoes and cucumber. Spring opens up meadows and pastures to support livestock, which are otherwise housed. The catchment area is rich in natural resources, and is being exploited by the mining of natural gas, coal, diamonds, tin ore, iron ore and gold. The mining activities threaten the water quality and quantity that reaches the delta, and jeopardise efforts made by wilderness groups to preserve the integrity of the nature reserve. A lack of concern for the environment typified the Soviet era. Sad to say, the laissez-faire capitalism that replaced it is no better.

Lake Baikal has its own story. This World Heritage-listed site is a fine example of the grandeur and beauty of nature. With a depth of 1700 m, it is the deepest and oldest lake on Earth. It holds some 23 000 km^3 of water, which is over a fifth of the world's unfrozen freshwater. To give you an idea of the enormity of the volume of water held in Lake Baikal, if the discharge from all the rivers in the world were to be combined, it would take them approximately one year to fill the lake. Impressive! The Baikal area is a treasure trove of natural resources, with biodiversity that is unmatched by any other lake in the world. It is home to an outstanding variety of endemic flora and fauna. This is of exceptional value to science, an attribute that earned it World Heritage status in 1996. Considerable damage has already been done. We wait in hope that Lake Baikal will retain forever its World Heritage status.

The Murray–Darling River system

Australia is a large country with a relatively small population (22 million). This is fortunate, as much of the country is arid or semi-arid and a large population is out of the question. Of the top 20 deserts in size in the world, Australia has three, four out of the top thirty. Water is precious in this country, although it is not a water-stressed country due to its small population. The Darling River is the longest river in Australia, rising from several headstreams in the Great Dividing Range in central and southern Queensland. Queensland is the second largest state in Australia (reaching from within a few km of Papua New Guinea in the tropics to the semi-tropical tourist town of the Gold Coast on the New South Wales border). The Darling River is joined by the Murray River, which flows west from its source in the Snowy Mountains in New South Wales to form the border between Victoria and New South Wales. Together these two rivers (known as the Murray–Darling) form the largest river system in Australia.

The Murray–Darling is 3750 km long and its basin covers over 1 million km^2. The two rivers and their tributaries (the major ones being the Murrumbidgee River and the Lachlan River) drain over an eighth of the Australian continent, including most of New South Wales, all the ACT, much of northern Victoria, southern Queensland, and parts of South Australia. Their combined waters are discharged into the Southern Ocean, east of Adelaide, after they flow through Lake Alexandrina.

The Murray–Darling has played a very significant role – probably more important than any other national feature – in the social, cultural and economic development of Australia. Aborigines have relied upon the catchment's natural resources for possibly 60 000 years. Fish and a variety of native flora and fauna found along its banks comprised the totality of the diet of people who populated the river environs until the Europeans came to settle in 1788. Over 10 000 traditional Aboriginal sites have been identified within the catchment, including rock paintings, ceremonial areas and archaeological deposits. Many of the original tribes still inhabit small parts of the catchment.

European exploration and settlement along the river began in 1824 when Hume and Hovell first saw the Murray River, followed by the Darling in 1829. Despite the irregular flow of the river determined by droughts and floods, it soon became an important transportation route, and agricultural development occurred on the extensive flood plains. When it rained they were flood plains, otherwise they were near-desert. Had these two rivers not existed, the Australian wool industry would have struggled to gain a foothold in the United Kingdom wool milling and, having achieved that, come to dominate the production of medium to fine wool worldwide. Until well after the Second World War Australia 'rode on the sheep's back', so important was wool to its economy. What is today the tiny town of Eucha on the Murray River in western Victoria was in the nineteenth century the busiest port in Australia. Wool travelled from far upland (and, obviously, inland) pastoral properties by small paddle-steamers to Eucha. It was then transported by larger paddle-steamers to the ocean, where it was again trans-shipped to be taken to the UK woollen mills.

Today, one-third of Australia's total output from natural resource-based industries is produced within the catchment, contributing approximately well over $20 billion per year to the Australian economy. Agriculture is the dominant activity, with over 50 per cent of the total Australian agricultural production coming from the Murray–Darling Basin. In the lower catchment, the Murray Irrigation Area is the centre of farm production. As the name implies, here the water is provided by irrigation drains. Major products are wheat, barley, oats, rice, beef and dairy cattle, wool, grapes and orchard fruits.

There are over 120 large dams on the Murray-Darling,[3] which irrigate approximately 1.5 million ha of crops and pastures as well as providing power to numerous hydro-electricity facilities. The catchment rivers also supply water for domestic and industrial use for the large number of towns (some tiny, some large) that are scattered along their banks. Parts of the river system are popular with tourists for recreational activities such as fishing and boating. In the lower reaches of the Murray a small tourism industry is based on the river boat excursions, some spending days slowly winding along the river. Adelaide and Canberra are the only large cities within the catchment, but numerous small towns (reaching far north into Queensland), have long existed on the banks of the main rivers that feed into the Darling River.

Within the relatively short period of European settlement, an enormous amount of damage has been caused by inappropriate farming and urban practices in the Murray–Darling basin. The over-use of water resources, particularly during prolonged drought, land clearing and pollution from agricultural runoff have greatly affected the health of the river system. Salinity and erosion have taken a toll on the water quality and on agricultural soils, with the occurrence of salt pans and water-logged paddocks. The construction of dams and reservoirs has significantly reduced natural flows, and algal blooms have severely altered the natural state of the waterways.

The management of the Murray–Darling basin has been and remains very fragmented and uncoordinated. The South Australian, Victorian, New South Wales and Queensland state governments have looked to their own interests, and the Commonwealth Government has been powerless to do much when conflicts between the states have occurred. In recognition of the need to manage the system as a whole, the Murray–Darling Commission was established in 1918 – a very long time ago. However, whatever has been done to date has been a failure – not yet dramatic, as much of the basin has managed to retain its agricultural productivity. However, there is a need today to buy back large quantities of water that have been allocated to farmers and reallocate these to environmental flows. Progress on addressing the environmental and economic health of the Murray–Darling Basin has been, and is, frustratingly slow. In a perfect world, the Murray–Darling catchment would have been a political entity. Overlap and duplication would not be problems, and a degree of fairness in water allocations would have occurred. A second-best solution is being sought, but entrenched property rights – including tenuous ones – mean that nothing worthwhile will be achieved without enormous political will.

The Volga River

The Volga is the longest river in Europe, and the principal waterway of western Russia. This is a Russian river we know much about. Starting in the Valdai Hills north-west of Moscow, it flows 3532 km across two-thirds of European Russia. Together with its 200-odd tributaries it drains an area of around 1.38 million km^2. The catchment has four distinct geographic zones – the dense, marshy forests of the upper reaches, the forest steppe near Saratov, the steppe of Volgograd, and the semi-desert lowlands near its mouth at the Caspian Sea.

The Volga catchment is home to over half the population of the Russian Republic. The Volga provided the means to the creation of Russia, and it is known as the 'cradle' of the Russian state. Today, the river is of immense economic and historical importance to Russia. Over 1000 years ago, in 980, the Swedish Viking Vladimir created the Rus (Viking for Russian) state. This chap was a real character. Originally a pagan, he converted to Byzantine Christianity because he considered it a more liberal religion than either Roman Catholicism or Islam. Importantly, his new religion allowed him to partake of alcoholic drinks, something to his liking. Such are the reasons to select a religion! Kiev was then capital of Rus(sia).

A favourable climate and the geography of the Volga made the establishment of Rus(sia) possible. Today, during the warmer months, the river is navigable for some 3220 km (over 90 per cent of its length), and together with its 70 navigable tributaries it carries more than half of the Russian inland freight, as well as a large number of passengers. What a wonderful blessing is the existence of a navigable river! There is no cost of construction and no maintenance (if you look after it), as there is for a highway.

Canals connect the Volga with other important waterways such as the Baltic Sea, and with the capital city of Moscow via the Moscow River. There are more than 900 ports and 550 industrial docks constructed along the banks of the Volga, including the substantial port cities of Tyer, Yaroslavl, Samara, Astrakhan and Volgograd. The cargo carried consists of construction and raw materials, petroleum, coal, foodstuffs, salt and agricultural machinery. Commercial and recreational fishing take place in the lower reaches of the river. Tourist cruises occur between the much-sort-after destinations on the tourist route. In terms of length these are not surpassed on any other inland waterways.

A total of eight reservoirs have been constructed along the Volga, and another three along the Kama River, one of its major tributaries. Collectively they produce 11 GW of power. As always and everywhere, reservoirs and dams have environmental implications, often major. In the case of the Volga, they have blocked access to the river for native anadromous fish species from the Caspian Sea. These are fish that used to spawn in the river. The impoundment of water has significantly reduced the discharge at the mouth, contributing to a steady decline in the level of the Caspian Sea. Industrial pollution, municipal waste, siltation and agricultural runoff have also severely altered the habitat of native fish species, which in turn has had a negative impact on the income from local fisheries. The Russian government has recently put in place tighter regulations in an attempt to reduce further damage to the river. The new Russian government is not going to be as environmentally derelict as the previous Soviet government and those that immediately followed. The health of the Volga can only improve and that can only be a good thing, for there is nothing more Russian than the Volga River.

The Indus River

Some rivers are readily associated with ancient history; the Indus is one. The Indus River begins its journey some 5500 m above sea level in south-west Tibet. It continues to flow 2900 km across India and the semi-arid plains of Pakistan to reach its mouth at the Arabian Sea. The catchment covers an area of 16 770 km^2, 80 per cent of which is controlled by India and Pakistan and the remainder by China and Afghanistan. The most notable tributaries arise from the Punjab plain, the 'Land of Five Rivers', a fertile section of land that is shared between Pakistan and India.

The Indus River Valley forms the cradle of the ancient Indus civilisation, one of the earliest known human societies. It flourished at the same time as the ancient civilisations of Crete, Mesopotamia and Bronze Age Egypt. The Indus culture was known as the Harappan civilisation. It was rich in foods (rice and wheat) and had the power of domesticated elephants to do any hard physical work. There was manufacture of cotton fabrics, and trade was well established with Persia and Afghanistan. Then in circa 2000 BC, the river course changed, agricultural fields were without water and abandoned, and as a direct result trade suffered. Such is the economic power of water!

Today, the Indus River basin remains one of the most populated catchments in the world with a density of 165 people per km^2. A large percentage lives in 11 major cities, including Islamabad, the capital of Pakistan, and Kabul, the capital of Afghanistan. The Indus is the lifeline of Pakistan, providing key water resources for its economy, and for this reason we will concentrate on Pakistan while acknowledging the importance of the Indus to India.

The soil in the valley is very rich in silt, and it is one of the most intensely farmed areas of the world. The large, and ever-growing, Pakistani population determines that. The Punjab plain accounts for most of the nation's agricultural production and employs between 50 and 60 per cent of the workforce. The main crops are rice, wheat, sugarcane, cotton and animal fodder. The cotton industry links to the spinning, weaving and textile manufacturing industries, and these help Pakistan earn much needed foreign exchange.

Productivity of the fertile plains has been boosted by the extensive, modern irrigation system that supplies water to an enormous 14 million ha area. However, as with every modern irrigation system, human ability to manage extraction and flows is far from perfect, and agriculture is not anywhere nearly as efficient as it could be. The river supplies water to a variety of heavy industries and provides the main supply of potable water in Pakistan.

The continuing rapid expansion of the Pakistani population, the very high illiteracy rate (at approximately 50 per cent), the lack of environmental expertise, and corruption have combined to keep Pakistan close to the bottom of the world's development rankings. There are high levels of poverty, food insecurity, and environmental degradation. Poorly managed irrigation systems have caused salination and water-logging of prime agricultural land. Dams have caused a reduction in freshwater discharges and the nutrient-rich sediment load that used to reach the delta. This has led to a reduction in the overall health of the flood plain and coastal environment. Of particular importance is the impact on the mangrove forests that provide habitat for fish, shrimp and migratory birds, which in turn supply local communities with a source of food and income. The future is bleak if the county's population cannot be brought under control, and the nation proves incapable of providing itself with effective, modern secular government. A large river counts for little in these circumstances.

The Ganges River

India, much touted as the next China, has a population that is quickly outstripping the Chinese and a business culture ready-made to trade with the English-speaking West. The country could one day supplant China in the development stakes. So far, India's economic growth is slow, except in Mumbai and other industrial–commercial centres. Agriculture still employs two-thirds of Indian workers – growing rice, tea, sugar, wheat, cotton, coffee, jute, oil seed, maize and pulses. The country's flood plains are the defining feature of this agricultural economy.

The 'Great River' of the plains of northern India, the 'Mother Ganges' (or Ganges Ma) flows 2520 km from the Himalayas (at Gangotri), south-east through to Bangladesh and into the Bay of Bengal at Kolkata (previously Calcutta). Although it is the shortest of the world's great rivers,[4] it is by far one of the most important. The Ganges flows through some of the most fertile land of the world. The river's religious significance exceeds that of any other water body. Who has not seen, if not in person then on television, the multitudes bathing themselves in the Ganges at religious festivals? The river is worshipped as a goddess. Its waters are deemed to grant immediate exemption from the trial and tribulations of reincarnation. Hundreds of millions of Indians worship the river, and most will make a visit once in their lifetime. Religious belief this may be, but at what cost to health? The river carries 300 times the level of fecal coliform bacteria acceptable to humans.

The Ganges catchment covers an area of approximately one million km^2. It contains one of the largest flood plains in the world, and is bordered by mountains and highlands on three sides. The Ganges basin (the most densely populated of any catchment in the world) supports a population of over 450 million people. This is in the order of one-third of the present Indian population. There are 11 large cities within the basin, including Kolkata and its many satellite towns which stretch some 80 km along the banks of the Ganges and its tributaries. A large percentage of the Ganges population lives in dire poverty, particularly those in rural areas. This has resulted in increased migration to the cities and large peri-urban slums, characterised by inadequate sanitation, substandard housing, no electricity and fresh water, and disgusting poverty. These slums are some of the worst in the world. On the other hand, the Indian billionaires are some of the richest people in the world. For all the talk of India being a 'socialist secular state', the opposite is true. Capitalism of sorts is dominant. The continued existence of the caste system (now unofficial) makes a mockery of the fraternity of human rights. Yet the country is far more democratic than most developing countries. Herein is its hope.

Historically the Ganges constituted the heartland of India and formed the cradle of its successive early civilisations. Its religious significance can be traced back to the earliest civilisation, and it remains the most important of the sacred rivers in India. It is a place of life and death, and a meeting spot for the rich and the poor when seeking solace. Hindus from all over the world make the pilgrimage to the Ganges River. It is believed to be the route to heaven, which is why the bodies or ashes of the dead are placed in it to be transported to 'the world of the ancestors', while the living bathe and drink its waters to gain forgiveness for their sins and attain salvation – regardless of the polluted state of the water.

Irrigation, together with the deposition of an enormous amount of silt, has allowed agriculture to thrive within the catchment. Crop growing currently accounts for over 70 per cent of land use. The western plain of the river has been cleared of its former grasslands and forests to make way for crops, which are irrigated by an extensive canal system. The major crops include rice, sugarcane, lentils, oil seeds, potatoes and wheat. Along the swamps and lakes that line the Ganges, and within the fertile delta, crops such as rice, legumes, chillies, mustard, sesame, sugarcane and jute are cultivated.

Unfortunately, with all the life the Ganges brings, pollution is a major by-product. The manufacturing industries situated along the banks of the Ganges release large amounts of chromium plus other dangerous chemicals into its water. Municipal waste, industrial pollution and agricultural runoff, together with the disposal of the dead, have combined to produce very poor water quality. So great is the load of human and industrial waste that it is threatening fish populations, dolphins and turtles. Some of the worst waterborne diseases known to humans, such as dysentery, hepatitis and cholera, are spread via the waters of the Ganges, yet this does not inhibit religious bathing. Faith prevails!

Irrigation and land clearing upstream are also a problem. Deforestation of the mountainsides has significantly reduced the Himalayas' capacity to absorb monsoon rains, greatly increasing the amount of eroded soil that is carried by floodwaters throughout the length of the Ganges. Large sections of the delta are now choked by silt, making them prone to flooding. In fact, only a small stretch of the delta remains covered with its original mangrove trees.

The Ganges River exits the sea in the Bay of Bengal, where a maze of canals, swamps and marshes constitute the delta in Bangladesh. Here the Ganges is joined by the Brahmaputra and Megha rivers to form the world's largest delta. All major rivers flowing through Bangladesh have their origins outside Bangladesh, and the quantity and quality of the water that enters the country is ultimately determined by its upstream users. This is a classical transboundary problem, and the sharing of the Ganges' resources has been a source of conflict. Rice growing is the major industry in Bangladesh, employing just under 90 per cent of the working population. Fishing is an important subsistence and cash industry in the delta.

The rapidly growing population in both India and Bangladesh puts an enormous strain on the natural resources. As more people migrate to the cities, demand for potable water is rapidly outstripping supply. Scarce water is dear water. This reinforces poverty and leads to further deterioration in socioeconomic conditions, which are already tragic for the poor of both countries.

The Tigris and Euphrates rivers

From their separate origins in eastern Turkey (Mt Ararat at 5165 m, the highest mountain in the country), the Tigris and Euphrates rivers flow south-west into the southern alluvial plains of Iraq. There they join and flow as the Shatt al-Arab, which empties into the Persian Gulf. The Euphrates is the longest river in this part of the world. It is formed by the union of the Kara and Murat rivers, and flows 2800 km through southern Turkey, Syria and Iraq. The Tigris is 1800 km in length and is fed by a number of small tributaries that drain much of eastern Turkey and Iraq. The lower sections of the combined Tigris–Euphrates River system form the 'cradle of civilisation' – Mesopotamia, literally meaning 'between two rivers'. Together with parts of Egypt watered by the Nile, Mesopotamia makes up what was known as the 'fertile crescent', an area of extreme importance in the development of agriculture followed by civilisation.

The first known civilisation in the world developed in Mesopotamia some 6000 years ago. The gods who provided for the people were Anu (the god of heaven), Enlil (the god of earth), and, not to be overlooked, Enki (the god of water). It is here that agriculture was originally developed in the order of 9000 to 10 000 years ago. The land adjacent to the rivers, the immediate floodplains, was fertile. Go beyond this area and the soil was hostile. The Mesopotamians solved this problem – in the short term – by constructing irrigation systems, dams and aqueducts. The long term is what we see today in modern Iraq – dirt, dust and desert.

The development of irrigation allowed farmers to increase their yield beyond what they could eat; the surplus was used to trade, and a new class of merchants, artisans, administrators and scholars slowly formed – hence, civilisation. The rivers and their tributaries allowed goods to be transported to and from surrounding areas, and Mesopotamia quickly developed into a major centre for trade and commerce. Wheat, barley, flax and wool produced in the valley were traded with people from the mountains for timber, limestone, gold, silver and copper. By the fourth millennium BC, the world's first known literature was developed.

The modern waterways of the Tigris and Euphrates rivers do not resemble the fertile flood plains of ancient times. This part of the Fertile Crescent gradually lost its initial agricultural productivity millennia ago. In more recent times, years of mismanaged agricultural activity, political turmoil and war since 2003, and the drainage and diversion of the watercourses, have taken a terrible toll on the health and productivity of the area. Marshlands have been destroyed, the rivers are clogged with silt, and the surrounding land has become barren due to severe desertification and salination. Only recently have the so-called Marsh Arabs commenced to rehabilitate this area.

The combined catchment area of the two rivers is 765 742 km^2, with a high population density today of 57 people per km^2. There are 28 large cities within the catchment, including the Iraqi capital of Baghdad, the ancient cities of Aleppo (Syria) and Kirkuk (Iraq), and the major sea port of Al Basrah at the mouth of the river in Iraq. Although oil revenues now play a major role in the Iraqi economy, the Twin Rivers remain a dominant force in the economic livelihood of all three countries within the catchment. Too shallow for navigation, the Tigris and Euphrates are used primarily as a source of water for agriculture. The major field crops in Iraq are wheat, barley, millet, rice, corn, sugarbeet, dates and sorghum. In Syria the mainstay is olives. Most of these crops are grown in semi-arid conditions and rely heavily on irrigation. The Euphrates Dam provides close to 100 per cent of Syria's domestic electricity as well as providing much needed irrigation water.

The sharing of the Euphrates water has been a point of political tension between Turkey, Syria and Iraq for many years and has become a greater source of agitation as the population of these countries continues to increase. Iraq has more water than Turkey, Syria has less, and all three want more. Water is so scarce in the Middle East that nations are threatening each other with war. Dam construction and water diversion will undoubtedly lead to conflicts in the future unless water sharing schemes are agreed upon quickly.

The recent construction of irrigation and hydroelectricity facilities under the South-eastern Anatolia Project in Turkey is a classic example of conflicting interests that arise when rivers cross national boundaries. While these dams have allowed Turkey to enhance the productivity and employment opportunities in rural areas, downstream in Syria and Iraq the decreased flow has affected the livelihood of their agricultural and energy sectors.

The Mekong River

From its headwaters in the Qinghai province of China near the Tibetan plateau, the Mekong River flows 4800 km to form the longest river in South-East Asia. Crossing through the Chinese province of Yunnan, the river goes on to form the border between Myanmar (Burma) and Laos, and most of the border between Laos and Thailand. It then flows across Cambodia and southern Vietnam, where it empties into the South China Sea near Ho Chi Minh City, draining a total area of 810 000 km^2. There are nine large cities within the catchment, including Vientiane, the capital of Laos, and Phnom Penh, the Cambodian capital, both of which sit on the banks of the river.

The Mekong catchment is home to over 60 million people (a small number in an Asia context) and approximately 100 different ethnic groups, making it one of the most culturally diverse regions in the world. The population density is very high throughout the basin, with the heaviest concentrations in the delta in Vietnam and in the dry, infertile Khorat

Plateau in north-east Thailand. Agriculture, along with fishing and forestry, are the key industries, in which 85 per cent of the people living in the catchment are employed. Extensive irrigation systems have allowed the area to produce a large amount of rice, as well as sugarcane, cotton, bananas, and various other fruits and vegetables.

With the exception of a small stretch along the Lao-Cambodian border, the entire length of the Mekong is navigable to smaller vessels, with sea-going vessels reaching as far north as Phnom Penh. There are 25 major ports along the river, including Chang Saen in northern Thailand, and Kampong Cham and Phnom Penh in Cambodia. Dams along the Mekong and its tributaries collectively produce 1600 MW of electricity, most of which is used to power cities and industry in China. In the past 10 years there have been numerous proposals for dams along both the upper and lower Mekong; however, the recognition of the potential adverse social, economic and environmental effects of such developments has prevented (or delayed) their construction.

The Mekong is second only to the Amazon in terms of biological diversity. It has a huge number of native plant and animal species, in particular fish, with over 1500 species native to the river. The catchment is rich in natural resources, including tin, iron ore, natural gas, potash, gemstones, gold, building materials and pharmaceuticals based on natural rainforest products. Despite the abundance of resources, the majority of the population in the catchment live in poverty. Fortunately for those living alongside or working the river, the Mekong is one of the most productive fisheries in the world. It not only supports a thriving commercial fishing industry, but also provides nutrition (in particular protein) for the majority unable to purchase food in the markets.

Population growth and economic growth have accelerated resource exploitation in the Mekong catchment. The river is an economic lifeline for the 60 million people from the six catchment countries. The water resource management issues and priorities differ in each of these countries, and the sharing of the Mekong resources has inspired both conflict and cooperation among them. The recent proposal from the Chinese to turn the Mekong into a major shipping route for oil is an example of possible conflicts of interest that exist between the various users within the catchment. The need to address trans-boundary water management issues was recognised by the United Nations, which established the Mekong River Commission with the task of promoting sustainable development in the catchment. China and Myanmar are not participating in the commission – illustrating the fact that it is not always the United States which thumbs its nose at the United Nations. Will China become the United States of Asia?

At this point the Mekong is the lifeblood of Laos and Cambodia, and a wonderful agricultural and fisheries resource for Vietnam. Approximately 70 per cent of the workforce in the delta is employed in agriculture. If you know these countries, try to imagine them without the Mekong River.

In conclusion

In a very short space we have attempted to provide insights into the world's major river basins. We have not provided any detail on how people live, what their values are, what their relationships are with the rest of the world. We have done little more than illustrate a very obvious point – civilisations have formed around great river basins; some rivers and civilisations have disappeared because of over-use of water, yet most remain today to play a crucial role in feeding large local populations and generating income through exports.

To conclude with a different perspective, consider the Netherlands. More than one-quarter of this nation is below sea level, protected by human-made dykes and coastal dunes. Flowing into and through the Netherlands are the Rhine River (from Germany) and the Maas River (from Belgium). The reclamation of land in Holland is one of those fascinating stories that have entered school-book histories and geographies. We all know the story from our school days of the boy with his finger in the hole of the dyke. The more pessimistic of us see this as a precursor to an unpredictable future fashioned by climate change.

7

Accounting for water

Peter Daniels

This chapter develops our understanding of environment–economic–social relationships. Here we will make progress towards explaining the concept and tools of water accounting. No natural resource is an 'island', and water is perhaps the most interdependent resource of all. All natural resources are intimately connected to others by complex and powerful ecological and other biophysical processes. For example, the forests that depend on the sun, water and soil also have marked effects on the conditions and services provided by water and soils. With industrialisation and population growth, the influence and impact of human (or 'anthropogenic') activity on natural cycles is a fundamental feature of our economies. When we combine the basic necessity for water with the need to understand resources as part of closely interconnected natural systems, there is little wonder that integrated water resource management has become a major focus and approach of modern environmental management and science.

Integrated water resource management is:

> *a process, which promotes the coordinated development and management of water, land and related resources in order to maximize the resultant economic and social welfare in an equitable manner without compromising the sustainability of vital ecosystems. (Global Water Partnership Technical Advisory Committee 2000, p. 22)*

Integrated water resource management is pursued in the broader notion of integrated catchment management. Not that long ago, water flows and use were studied in a piecemeal fashion in ad hoc regions, often geopolitical in nature. In the new bioregional and integrated approach, the catchment as a water-flow network system provides the basis for capturing and understanding all major ecosystem elements as well as related human, economic and social activities, impacts and values.

The basis of integrated measurement and management

This central role of water in the integrated study and management of natural resources reflects the importance of the substance. Water underpins and connects all forms of life. The strength of this connection means that identifying, understanding and measuring the main stocks, flows and processes associated with water is vital for effective human–environment interaction. Understanding and measuring are close partners in decision-making based on science. The endeavour is to understand and measure the links between water in nature and its use by, and impact upon, society. An integration of the natural and social sciences comprises an ideal laboratory for interdisciplinary approaches that are necessary to effectively manage water on a sustainable basis (in terms of all three criteria – economic, social and ecological).

In the past, the study of water as an environmental resource was dominated by hydrology. This emphasis followed on from an assumption that the water cycle operated primarily outside of the human domain. This, of course is not true. Today, the scale of interaction between human activity and the 'natural' water cycle is so significant that it is not possible to study water effectively just as a hydrological phenomenon directed by natural systems. The linkages between the water cycle, other natural fluxes and resource conditions, and human activities, are pervasive. They include interconnected aspects such as the extent and type of vegetation and land uses, water abstractions and translocation, and energy production and use. These have links to microclimate, rainfall levels and variability, temperature, evapotranspiration, runoff, and surface and groundwater flows. In a nutshell, the land cover and water stocks and flows of most catchments are significantly affected by human activities. Water provides four major functions for humanity: as ecosystem services within nature (for example, the role of moisture in creating productive soil conditions); as an economic input, to be combined with fertiliser in agriculture; as a direct consumption product; and as a residual or waste product (polluted water) or 'loss' removed from the economic system back to nature (United Nations 2003).

Hence, a major prerequisite for water management is to identify and quantify the stocks and flows of water for both natural and human use (water captured and impounded), with the catchment as the system boundary. This goes far beyond the specialised and partial approaches of hydrological modelling that have prevailed in the past, towards the development of much more comprehensive empirical water accounts that map the significant stocks and flows within and between both nature and society.

Water accounting and budgets

'Water accounting' is a general term for the systematic collection and compilation of empirical information measuring physical volumes of water flows and water stocks within a defined area, usually a catchment. Accounting exercises, be they biophysical or monetary, tend to involve the principle of balance. With a given system boundary, flows into the system must equal flows out of the system, and any difference must be attributed to change in the stock within the system. For most biophysical phenomena, mass balances must exist for predefined systems in accordance with the First Law of Thermodynamics (Ayres 1978).

The idea of balancing accounts for water supply, use and stock levels provides the cradle for the notion of water budgets. In a general sense, water budgets are a summary technique for balancing inputs and outputs through an agreed system. The impetus for creating budgets is the desire to use water prudently. As with a household's budget, it involves knowing what 'income' (water supplies) one has access to, and how that income is 'spent' (water use) and whether this pattern of 'expenditure' is what we want, or is giving us the best outcome.

Leaman (2007, p. 7) makes the point that 'the water budget at a place is an appraisal of inputs and outputs in the same way we would describe our financial affairs'. While this is certainly true, there are some significant differences. Most of us expect a constant pay packet, week-in, week-out. The exceptions to this are (mainly) small business people who, more so than anyone else, are subject to the vagaries of the market. We cannot expect the same 'pay packet' of rainfall on a weekly basis. Rainfall can be as irregular as the income of a small business, which reflects seasonal fluctuations, changes in interest rates and the exchange rate, and a number of other external factors. Farmers are the classic case where both natural and economic dynamics combine to lead to great uncertainty. Rainfall is one of the natural variables facing a farmer; other natural variables are extreme weather events, while economic ones involve changes to demand for the farmed products. The 'income' of our water budget is rain or snow. We will see later just how variable that can be in our case study catchment.

With water budgeting as a management tool the aim is to ensure that businesses or the community allocate water first to those uses considered essential, followed by those giving the highest returns. Is more income earned from applying a unit of water to a hectare of dairy farm land or an orchard? This is the type of empirical question addressed in the water budgeting exercise. It follows that budgeting is intimately tied to the concept of 'efficiency' in economics. The conversion of physical quantities such as the volume of water to dollar values is necessary if we are to measure its net benefits or value.

Under the principles of sustainable development, our use of water has to meet specific criteria, one being the marriage of ecological function to economic outcomes. This requires us to measure the benefits and/or costs of a large range of feasible interventions in the water cycle. Hydrology and related natural science disciplines have a supply-side emphasis on water processes.[1] When economics and natural sciences are combined, we have a complete framework of water accounting. The achievement of this goal is only possible if we can obtain sufficient ecological and economic data. Given that we are not used to thinking of catchments in this holistic way, we are likely to discover that those who could routinely gather such data do not.

Our goals for water accounting in Little Swanport

Our practice of water accounting is undertaken to better understand the human-influenced water cycle of a predominantly rural catchment – in our case, the Little Swanport in Tasmania, Australia. The principles we apply could be used to study any catchment, and we hope they are applied widely and any omissions or errors in our work can be identified with the objective of improving water accounting. The Little Swanport catchment has been subject to several detailed studies of its hydrological characteristics.[2] While it is a minor catchment (some would argue tiny) in terms of overall national geographic and economic significance, there is great interest in water allocation and use in the catchment. The classical upstream-downstream competition for both water quantity and quality surfaced in the catchment community some years ago when an upstream land-holder applied to the government to construct a large dam. No large dams exist in the catchment; rather there are hundreds of small water-holes on the farms. At the very bottom of the catchment, in its estuary, there are oyster farmers who have a concern about the availability and quality of water for their production.

The models we developed and outline in this book are akin to a water balance model, with scenario assessment; the focus is on the application of several alternative methods that detail the nature of economic activity and associated water demands in the catchment. The modelling is unique in the extent to which it links hydrological and economic activities within an integrated catchment context. We assess the catchment economy's water demands in relation to total water availability and flows and, at least in a preliminary way, the value of output from economic production in the catchment.

However, the research reported here in is not the end of the matter. It does not intend to present a comprehensive set of water accounts. We were not able in the time available to obtain sufficient data – in fact we would have needed to do extensive groundwater surveys to complete the catchment's water cycle. Rather, we developed an exploratory water cycle account framework as a foundation for those who follow. Our framework provides data to map the structure and flows of the Little Swanport water cycle and reveal the groundwater data gap that is presently a limiting factor.

Water accounting advances

Globally, there has been rapid growth in interest and research in water accounting. Resource accounting is being pursued in recognition of the fact that if we don't measure what we have and how it gets used there is little chance of managing resources sustainably. Water is of particular interest in the accounting of environmental resources. Its fundamental life support and economic roles have made it subject to a great deal of tension – ranging from economic competition to violent conflict and war. Increased uncertainty about security of supply with climate change has undoubtedly increased the perceived importance of water.

While these issues are important for all nations, the effective management of water is even more critical in regions where water is scarce or supply is highly variable. Hence, there is little wonder that Australia has invested substantial resources (involving tens of billions of dollars in recent years) in improving water security and efficiency and has taken a leading role in related international efforts. The guiding principle for these efforts is the National Water Initiative (NWI) – a 2004 intergovernmental agreement across most of the nation's states and territories – being administered by the National Water Commission (NWC). The NWI is planned as an appropriate response to

> *the continuing national imperative to increase the productivity and efficiency of Australia's water use, the need to service rural and urban communities, and to ensure the health of river and groundwater systems by establishing clear pathways to return all systems to environmentally sustainable levels of extraction. The objective of the Parties in implementing this Agreement is to provide greater certainty for investment and the environment, and underpin the capacity of Australia's water management regimes to deal with change responsively and fairly. (COAG 2004)*

The implementation of comprehensive, detailed and accurate water accounts is recognised as a vital component of the implementation of the NWI.

At the international level, water accounting is primarily developing under the auspices of the United Nations Statistical Division's *System of Environmental and Economic Accounts for Water* (SEEAW) (United Nations 2006). This is part of the more generalised System of Environmental and Economic Accounts (SEEA) drafted in 1993. The SEEA is directed towards environmental–economic accounts – that is, integration of environmental resource stocks and flows and specific economic activities rather than the direct modelling or measurement of resource flows within nature. As in the case of integrated catchment management, new approaches are responding to the need to link natural and human functions.

The aim of SEEA is to develop and compile widely accepted and conceptually robust physical 'satellite' accounts to complement traditional national economic accounts (Australian Bureau of Statistics 2007). A range of natural resources (natural capital) that are not represented, or are inadequately represented, in the standard economic accounts were identified to be subject to satellite accounting. Water is one of those resources. The long-term intention is to convert the volumetric accounts to monetary ones, and hence allow budgeting of water use in economic terms. At this stage in their development, scientists are struggling to construct accurate physical accounts: how much water is in the catchment, how much is in groundwater and – too often – how much is extracted and subject to evaporation as well as devoted to 'economic' use. As a digression, we repeat our frustration with the lack of research into groundwater in the Little Swanport catchment and the fact there is no monitoring of water use on the farms. Governments, who generally 'own' and control water, have not put in place, in far too many places, the means of measuring its use. If the physical accounting is problematic, then so too must be the economic accounting. However, we cannot wait until the gaps are filled as decisions on water allocations are made daily – some will be the wrong decisions, while others will be the correct ones, but whatever the case our ignorance means that luck will play a significant role. This is not a good position to be in with something as valuable as water.

Australia's persistent and severe drought conditions, and growth in water demand, can help explain the proliferation in efforts towards water management schemes and data gathering.[3] The urgency of the task has led to considerable divergence in the evolution of conceptual and data frameworks. Broadly speaking, a divide has developed between on the one hand analytic approaches emphasising hydrological aspects and physical supply and volumes or flows of 'natural' water (with some land use factors included) and on the other an emphasis on economic supply and use as adopted in the United Nations and related Australian Bureau of Statistics (ABS) accounts (e.g. Australian Bureau of Statistics 2006).

The unfortunate divide is closing with growing recognition of the need to understand and measure all biophysical and human aspects of the water cycle. Australia is currently working towards a set of National Water Accounts that it is hoped should combine the coverage of, and help link, both approaches: those leaning towards physical water cycle balance and auditing (such as the NWC's audits and 'common chart' of water accounts) and the economic focus on supply and use (by sector or industry) advocated by the United Nations and the ABS.

This integration, or at least explicit linking, would produce a very useful system of accounts and meet other key goals of the NWI.[4] The case study we report in this book aims to contribute to this process with a specific focus upon the need for and approach to integration of economic and environmental aspects of the water cycle.

Uses for water accounts

To begin creating useful water accounts requires thorough research and understanding of the entire water cycle's components, structure and processes. Understanding the system, and measuring it correctly, is the sound scientific and informational basis for effective management of water. Without knowledge of the iterative, complex relations in the stocks and flows of the water cycle, it is simply not possible to assess or predict intended or unintended changes that might be associated with climate, land use and cover, socio-demographics and the myriad of other natural and human conditions and influences. Hence, understanding depends on measurement. Most of the potential benefits of 'getting things right' with water accounts (see the list below) are specific instances of this basic proposition.

Naturally, the understanding and data provided by laudable water accounts are integral to the NWI and its key aim of increasing the productivity and efficiency of Australia's water use.

While many of these factors are interrelated, the benefits of water accounts include those outlined in Box 7.1.

Box 7.1: The benefits of water accounts

- The provision of detailed information on physical water use or demand, classified according to standard industry sectors. This includes changes over time and can be extended with input-output and life cycle analyses to identify virtual or embodied water use for final goods and services.
- Enhanced understanding of the structure and operation of natural processes governing water availability and cycling through nature. These 'supply-side' measures form the basis for determining environmental water requirements and sustainable yield levels available for human appropriation.
- The identification and quantification of all relevant linkages between natural aspects of the water cycle and human activity. This includes specific physical water demands and appropriations as well as more general overall human supply and demand effects on the flows and stocks of the water cycle. Focusing on the nexus between demand and supply for the resource provides the context for establishing water balance (or imbalance) conditions at any given point in time. It includes the assessment of the scale of human intervention in the water cycle and water quality outcomes if the relationships between human use (or land use) and natural processes are to be understood.
- Tracking and mapping the flows of water into, through and out of the economy will help identify and address water losses, leakages and wasteful human impacts. This capability is enhanced when mass balance approaches are used in water accounting, whereby discrepancies in stocks and flows can be validated and balanced to reveal losses (or data shortcomings). Areas for targeting technological and operational efficiency gains can be revealed and impacts monitored with water account data.
- The analysis of economy-wide water productivity so that those activities that provide the greatest economic return to society from water use can be identified. Ideally, the full costs and benefits of associated outputs and water supply (or use) should be known. This process facilitates water budgeting and 'social efficiency', where society gets the most welfare out of defined or limited water resources (over the longer term in accordance with sustainability principles). Commencing with the existing uses (including extractions, diversions and storages) to obtain the (marginal) value of water, we seek to determine the overall economic gain or loss resulting from small reallocations of water – take a few litres from in-house use and apply those litres to the farm; take a few litres from growing potatoes and apply those litres to growing grapes. These are what economists call changes 'at the margin'. Note that economists do not attempt to estimate the value of all the water in the world or in a catchment. They seek to better societal outcomes by reallocating water so that it is used in its 'highest value' use; that is, where a unit of water produces its greatest value-added.
- Identifying water productivity requires the preparation of 'hybrid' water accounts where, within specific industries or sectors, physical resource use is matched to that sector's dollar values of output or value-added. This provides a very powerful (but complex) analytical tool that can be applied to encourage the more efficient use of water so that it is allocated where it adds the most value across sectors, regions, and communities (Bain 2008). As noted by the ABS (2007, p. 2), 'linking monetary and physical water accounts provides information useful for determining efficient water

allocation, achieving cost recovery for water infrastructure assets and analysing trade-offs between alternative water and economic policies.' Water pricing and trading are typically integral to implementing such change (and for these to be beneficial tools we need accurate and detailed water accounts).
- Water markets, trading and pricing – water accounts are also critical for identifying and attaching monetary values to those cost and benefits of water supply and use that are not typically covered or recovered in market transactions. There are many such external effects. Water has common property attributes and, historically, water has been publicly provided so that those who use the resource do not pay the full cost of its supply and use. Externalities and unpriced effects permeate water resources use. These market imperfections have to be corrected.
- Water accounts can help assess how continued economic growth or anticipated economic and population change in different regions will affect the water balance, services provided by water and sustainability outcomes in general (Vardon *et al.* 2007). These include variable growth rates in different economic sectors (such as housing, resource-based industry, and power generation). Similarly, water accounts are pivotal in determining the sectoral or regional economic implications of government water policy and exogenous economic influences, such as technological advances.
- When comprehensive water accounts are consistent and connected to other environmental resource accounts, it is possible to assess and work to optimise environmental resource use across a wide range of interrelated areas (such as linking water to energy and greenhouse gas emissions). This is a key aspect of international environmental–economic accounting, which is in its infancy but generating much attention due to the need to comprehend the inter-sectoral complexity of climate change.
- Water accounts that adopt system-wide material flow or mass balance frameworks also help reveal critical missing information (or what additional information would be most useful in effective integrated catchment management, and how, where and when it might be collected).

Overall, it can be seen that the water accounting process is a prerequisite for any type of modelling capability of human–environment interactions. The ideal or aim is not just the sustainable use of water resources but actually increasing the social efficiency with which water is used. This presents the possibility of improving economic (and hopefully social) outcomes under a sustainable water management regime.

Most of the benefits of water accounts described above apply in the context of our case of the Little Swanport catchment. Only with a thorough understanding of water in an accounting framework can empirically-based decisions be made to improve water allocation and use. The Tasmanian Department of Primary Industries, Water and Environment (DPIWE) identified the following issues in the context of developing a water management plan for the catchment – and with an eye to the concerns raised when the large upper-catchment dam was proposed (Resource Planning and Development Commission 2006):

- potential effects of a proposed irrigation storage on river flows
- potential effects of irrigation water usage on catchment water quality
- perceived or real lack of reliable information about the sustainable yield of catchment streams.

Catchment-based water accounts

We are all connected through catchments.

A catchment is a web of life. The life it supports is interconnected, meaning every creature and plant depends on other creatures and plants in the catchment for sustenance. If the whole catchment is like our bodies, then rivers are like our veins – coursing with life and crucial to sustaining it.

We all live downstream from someone, and upstream from someone else. (International Rivers 2008)

Catchments are special areas of the Earth's surface. The land within their boundaries has many shared features and connections based on the simple fact of proximity. However, this spatial domain has numerous and very strong interconnections and pervasive interdependence based on its central binding essence – the collection and flow of surface water. Water, together with air and soil, are the primary inputs of life and, powered by gravitational and solar forces, the water cycle's terrestrial domain is centred upon catchments.

Catchments are defined as geographic areas where water enters the system as rain or other forms of precipitation and then flows according to the topography to one or more defined endpoints or outflows from the system boundary (usually a major water body such as a reservoir, lake, estuary or bay, or direct to the ocean). In a catchment, water flows under gravity from precipitation through terrain sloping down to some lowest elevation endpoint. A catchment's exact boundary can be very difficult to define, particularly because groundwater systems can differ from the surface flows that topology determines. However, a catchment does have a boundary, its 'watershed', that marks the divide between water flowing into one catchment or to another. It is a matter of establishing its precise boundary, and this is where research on groundwater flows is a prerequisite for an accurate definition of a catchment.

We can think of catchments at various scales, with large catchments comprising a hierarchy of streams (rivers, rivulets) and the lowest-order streams having no tributaries. A feature of streams is their gain or loss of water to groundwater storage. A stream can gain water in one reach and lose it in another. Groundwater tends to provide a stabilising effect upon the variability of surface water. 'Groundwater forms the great balancing fly wheel which keeps the water cycle in action and healthy … Its slow and steady processing keeps the atmosphere and marine elements of the water cycle stable and slows down the reactions which may occur in times of climate change or abnormal demand' (Leaman 2005, p. 11). Unless we can measure this, our water budgeting attempts of a catchment will be inaccurate.

It is the flow of life-sustaining surface water that connects and unites the catchment as an organic, evolving system. It is the nexus of land, water and human society. Human activities not only depend upon this relationship but also have a marked impact on its operation. All forms of agriculture and many industrial activities (such as manufacturing and power generation) as well as urban residential settlements are bound up with water cycle catchments. The attendant issues and problems for society (such as drought and sustainable water management) are best analysed within a catchment context.

The societal and economic linkages and topographic characterisation of catchments (which determine land use) produce a sense of place and community and help maintain the vital, but poorly understood, third pillar of sustainability – social capital. If there is one central feature now recognised as critical to sustainability and welfare, it is the notion of interdependence, and water is the great connector. If catchment dwellers – and water users in a catchment – have a strong sense of place and vigorous social capital, they are likely to understand that what happens upstream determines what can be done downstream, and will work to ensure the classical economic externality of downstream pollution and water scarcity does not occur. Of course, we cannot take it as given that such strong social capital exists.

The recognition of fundamental connections in catchments is aligned with the new, but very influential, science of ecological economics. This rapidly growing trans-discipline focuses on the need for integrated studies of human economies and society, firmly placed within the constraints or carrying capacity of the biophysical environment in which they are embedded. In particular, socio-ecological economics (to further broaden its title) covers all social, economic and ecological dimensions that are so important for integrated catchment management if it is to be focused on sustainability.

In accordance with the First Law of Thermodynamics and the mass balance property of biophysical systems, water is never lost or consumed, it simply moves across the system boundary or flows between system components or is stored within system stocks. Human influence only redirects some of these flows and stocks, but it may hide some of the system transfers in the export and import of water embodied in goods or via transport across catchment boundaries. In a large number of the world's catchment water cycles, human production and consumption have significant effects on both water flow quantity, mainly by storing and abstracting water from surface flows and land surface changes, and on water quality, by changes in land use, irrigation–soil flows, application of fertiliser and chemicals, livestock waste, and urban and industrial emissions.

The human uses of water (call them 'extractions') divert it from immediate direct contact with the environment. The routes from rain or snow to the environment are no longer unimpeded when there are extractions. Not long ago, when human numbers were few and industry and agriculture were basic, most water flow was from nature to nature. Today the route can be via rainwater tanks, dams, irrigation channels and, less obviously, into cultivated fields and plantation forestry, rather than on to natural grasslands and forests. The diversions and what happens to the diverted water alter the pace of the flows and the quality of water as it comes into contact with the wide range of pesticides, herbicides, chemical fertilisers, industrial effluents and human and domesticated animal wastes.

The natural state of a catchment?

What we can't do – although some water scientists and environmentalists would like to – is describe a region's water regime, and budget, in a state without humans (a pre-human world). This hypothetical world is what some, rather simplistically, call 'the natural state'. We don't know what this was. We can't recreate it with any degree of precision. We are very much left with working with what we have today. That does not mean that we cannot – or should not – restore water quality and environment flows for the betterment of the natural environment and downstream industries and residents. These things we can do, and have done very successfully in some cases. In applying economic analysis to water use, the starting point is the existing water situation – usually at a local level, ideally at a catchment level.

In summary, catchments are very special, and they thoroughly deserve to be studied as entities and systems or hubs of intensive interconnections based on water processes and closely related land, ecological and human factors. Each catchment is unique, with its characteristic profile of natural and socio-economic features and its relative significance within a part of the globe. This is the theoretical background to empirical analysis of catchments.

8

An introduction to the Little Swanport catchment

Tor Hundloe

From here on, the book deals in considerable detail with one small catchment. What goes on in the catchment is described, places of interest are identified and so are some individuals. Our research took us deep into the lives of people in a small catchment. Of course, we cannot, and do not, delve far into personal matters where they must remain personal. Our interests are the social, economic and environmental relationship in a tiny catchment. Historical facts and matters of community interest that residents shared with us, and are willing to share with the reader, are here to be read. When we come to discuss economic matters we report only averages, aggregates and the general situation. This is a promise all professional researchers make and keep.

In many respects the Little Swanport catchment could be any catchment in the developed world. We would undertake the same types of studies, we could use the same approach and tools to study any catchment, and we would expect similar types of associations between the use of water and economic and social relationships. It is this interdependence of humans and the environment, as we discussed in the preceding chapters, which we would expect to be similar – not in detail of course, but in principle. Scientific investigation seeks to find generalisations, principles if you like. We are not suggesting that the Little Swanport catchment is identical in a hydrological, geographical, demographic, economic or social context to anything that exists elsewhere. First, our catchment is very small, with only a few people living in it. There are no towns, let alone cities, within its boundaries. There are no dams as such, but rather a multitude of small watering-holes for sheep and cattle, and limited irrigation from the river and its tributaries. It has a myriad of small rivulets feeding into the main river, and hence gives the impression that it could exist as a catchment of a much grander scale. What makes the story of the Little Swanport catchment compelling is the level of detail available to us. Its small size and population are what make it possible to dig for facts that would not be possible (without enormous resources) in analysing a larger catchment.[1]

In comparison to our catchment, consider the size and complexity of the Murray–Darling basin, with its contributing rivers having their source in central Queensland, flowing through New South Wales and the ACT, and the Murray River itself forming part of the border between New South Wales and Victoria and eventually entering the sea near Adelaide in South Australia. This is a big system of a very big country. Very large rural properties extract water from the Murray–Darling system; numerous towns and even good-sized cities are built on the banks of its rivers; and there is the nation's largest hydro-electric scheme in the headwaters of a major tributary, the Snowy River. Our catchment – as the reader will soon find out – is in comparison to the Murray–Darling a mere baby.

It is now appropriate to step back from our analysis of catchments, water accounting and other matters pertaining to our investigations. To gain an appreciation of how we went about delving into and describing the human–environment relationship in our catchment, I need to explain who we are and what we do.

Our team and what we do

I need to explain how and why a small team of researchers have come to spend years studying one small catchment. Before that, allow me to introduce the team. It is a long way into the book to do this, yet it has not been necessary until now. Now that we are we about to delve into the creeks and gullies of the Little Swanport catchment, who are we?

The team leader is Christine Crawford, a senior scientist from a specialist unit of the University of Tasmania. There are not many universities further south in the world than the University of Tasmania. The southernmost university campus – not a central university – is in the small city of Ushuaia in Argentina. It is at 54°48′S whereas Hobart is 42°55′S. Christine is attached to the Tasmanian Aquaculture and Fisheries Institute. She is an ecologist specialising in estuaries, and recently won recognition for her research with a prestigious award from the university. Working closely with Christine is Jeff Ross, another estuarine ecologist. High-level estuarine expertise is a key requirement for our research. Supporting Christine and Jeff, with fortnightly trips to Little Swanport to collect water samples and bimonthly for oysters, was Sam Foster, their ever reliable offsider.

The remainder of the team work with me. Our research skills are in environmental science and environmental economics. Peter Daniels, who is a senior lecturer in the Environment School, Griffith University (in Brisbane), is an expert in the economics of water balance modelling. Understanding what happens to the water after it enters a catchment via precipitation is a fundamental requirement if we are to understand human–environment interdependencies.

Over the extent of the study period I have had three research assistants. These are keen young workers who do much of the on-the-ground digging for data and, once discovered and deciphered, type the information into text. Originally there was Tara Cully, who was enticed away by another research challenge before the study was completed. Her initial research on the social structure – the social capital – in the catchment provided a crucial foundation for what the rest of us have done since Tara left the team.

Michelle Wenner replaced Tara. Michelle has degrees in law and environmental science, and is a budding environmental historian. Environmental history is becoming an important discipline as the environment comes to play a central role in twenty-first century society, and there is much to learn from an investigation of historical facts. Yet environmental history is not taught as a specialised subject. As a consequence, Michelle's research skills are being learned by the most demanding method – experience.

The other member of my team was Amy White. Amy's original training was in ecotourism, from where she made a sharp turn to nutritional science. Both these (disparate) skills proved to be essential for our research. Ecotourism, which is a stricter form of nature-based tourism, is a significant industry in Australia, and in Tasmania in particular as much of the state is in its natural condition, is beautiful, and is accessible to the adventurous trekker, white-water rafter, or sedate sightseer. One cannot analyse the human values of the Little Swanport catchment without asking: what is its ecotourism potential? Amy's other skill, nutritional science, is a major help in better understanding farming and aquaculture in the catchment. Food, whether from fat lambs, potatoes or oysters, is produced because consumers purchase these foodstuffs. A variety of factors underlie human preferences for food, and the changes they make in preferences as time goes by. This is where nutritional science can help. Near the end of the project, Loren Angel joined the team and was extremely helpful in tidying up the manuscript.

This leaves me as the remaining member of the team. When I commenced this project I was Professor of Environmental Management at the University of Queensland. After a 10-year stint, over-burdened with teaching (which is a love of mine), I resigned and returned to my previous home, the Environment School at Griffith University, on a 3-year assignment. I had been there previously from 1976 to 1990. I then joined up with Bond University as one of its two Foundation Professor in Environmental Science and Management. My first discipline is environmental economics; however over the years I have added knowledge in ecology and ethics and I am a member of the International Society for Ecological Economists.

As with most major research projects funded jointly by government and industry, our research had a steering committee. The majority of its members were drawn from the local community. It was chaired by Professor Ted Lefroy from the University of Tasmania. Ted suggested the title of this book. Various community representatives, including pastoralists, oyster farmers and elected local government politicians, were on our steering committee. Individually and collectively they were a sound source of advice and encouragement.

What research did we do in the catchment and, just as importantly, why? Why this small catchment, one where there is no inter-jurisdictional conflict over access to water? It is a small part of a small Australian state, and hence the inter-jurisdictional conflicts we witness elsewhere over water allocations are not going to be evident. The catchment does sit in two shires (local government areas), but local governments have no sovereign rights to land or water in Australia. These matters are the responsibility of the states. There was potential conflict between users of water, and this was the catalyst for our research.

The Little Swanport catchment is approximately 87 600 ha and is situated on the central east coast of the Australian state of Tasmania.

There is, however, a small problem to deal with in defining the Little Swanport catchment. Two official definitions and separate boundaries for the catchment exist. One is purely a political construct used for land-use planning purposes. We use the physical geography of the catchment (river basin), while some parts of the Tasmanian Government use the larger political boundary.

The physical catchment area is determined by the Little Swanport River, around 61 km in length and having five main tributaries. This drains an area of approximately 600 km^2. There is another much smaller river flowing into the same estuary as the Little Swanport River. It is the Ravensdale Rivulet. Given that the water quantity flows and water quality of the estuary are important factors, because this is where the oyster-farming occurs, it is reasonable to consider all flows into the estuary. Together, the Little Swanport and the Ravensdale Rivulet contribute approximately 94 per cent of freshwater inputs to the estuary (Resource Planning and Development Commission 2006).

It is more difficult to justify why the Buxton River and Lisdillon Rivulet, some distance to the north-east and not in any way linked to the Little Swanport River or flowing into the same estuary, should be included in an official definition of the catchment. However, they are included in the 'management area' defined by the Tasmanian Government as the Little Swanport catchment. By including these watercourses, the 'catchment' increases to around 890 km^2. We need not be worried about this bit of bureaucratic interference with real world geography, as it does not affect our analysis.

The catchment is located in eastern Tasmania in a relatively dry region in the rain-shadow of the much more moist western side of the island. Rainfall in our region varies from around 800 mm in the higher elevations (at around 800 m above sea level) down to around 600 mm on the coast to the east. It is here, on the east coast, that Little Swanport estuary opens to Great Oyster Bay and the Tasman Sea. We are writing of averages when discussing rainfall. Like most of Australia, rainfall is variable and unpredictable.

The Little Swanport River has its source in the area of Inglewood Hill in the north-west of the catchment. The major tributaries of the river are Nutting Garden Rivulet, Crichton Creek, Eastern Marshes Rivulet, Pine Rivulet, Green Tier Creek and Pepper Creek. There are numerous other rivulets, a virtual set of octopus tentacles converging into the larger tributaries and the latter finding their way – ever downhill – into the main river. To complete the picture, there is the water that enters the catchment as rain (and occasionally snow) and percolates into the groundwater system before reappearing downstream as surface water. Of this we know virtually nothing, as it has not been established that there is a subterranean river and hence we are in the realm of speculation.

Eons ago, the relationship between water and the soil (and the plants that grew in the soil) brought to life the Little Swanport catchment. It remained unknown and untouched by humans until some time between 35 000 and 60 000 years ago when the people known in Australia as Aborigines reached Tasmania. At this time the island state of Tasmania was joined to the mainland. A major ice age had dropped sea levels to allow this. When the thaw, which commenced about 15 000 years ago, had finished, Tasmania had become a separate island. This is, as we tell far too briefly, the early known history of our study area.

A little history

The Aborigines, who came to Australia from south Asia via what is now Indonesia and Papua New Guinea, scattered over the immense continent of Australia, and those who reached Tasmania were to find themselves alone in the world when the island separated from the mainland. Alone! For those of us who regularly meet with people from the four corners of the globe, imagining the millennia of isolation of the Tasmanian Aborigines is near impossible. There were very few of them, probably in the order of 5000 (maybe 7000). It is estimated that about one-tenth (600–800) of them lived in the Little Swanport catchment, coming together for major feasts of oysters and birds' eggs at the mouth of the Little Swanport River. It must have been a productive environment for such a small coastal area. The large middens that are evident to those who know where to look on the southern bank of the estuary suggest this was the case.

After the Europeans came and the settlement grew, proto-villages developed within the catchment. Travel was arduous and slow, by horse and dray, or bullock cart if the load was heavy. The first settlers had to 'import' their needs, from farm tools to bed linen and bags of flour. These came from Hobart or the closer, smaller village settlements. Obviously, the purchase and transport of flour and nails (to nominate two regular needs of the early settlers) was much easier if there were local suppliers. As a matter of history, no major town formed within the boundaries of the catchment, although Oatlands was not far outside of its north-western boundary. It remains the case today that no town is inside the catchment.

The troubled waters of the Little Swanport

The catchment, its people and its economy would have gone unnoticed by the outside world had not the expansion of agriculture, and the prospect of irrigated agriculture, arisen. In the early part of the first decade of the twenty-first century, there was a proposal to build a relatively large dam (large in the context of the small catchment) in the upper parts of the Little Swanport catchment. If constructed, it was thought that the result would have been a significant increase in the value of farm production based on the irrigation. This would have been a significant land use change for a catchment with no dams except for small farm water-holes.

When knowledge of the proposed dam spread among the community, questions arose in the minds of downstream catchment users of water, in particular the oyster farmers. In particular, questions that scientific analysis should have been able to resolve were identified. There was an ecological question. What effects would the proposed dam, or any significant additional extraction of water in the catchment, have on the health of the estuary, on oyster farming downstream or, for that matter, on other downstream interests? To what extent did the profit of the oyster farmers depend on the existing quality and quantity of water flowing downstream into the estuary, where the river water mixed with the tidal flushes from the adjacent Great Oyster Bay? What we were dealing with here was the typical water-use conflict in catchments – what happens upstream has the potential to impact on what happens downstream. If there is no market mechanism that allows for trade between downstream and upstream water users, we have a real problem.

The catchment is by no means a pristine, natural environment. The upper reaches of the river have been farmed for about 150 years and the coastal area settled for about 180 years. Today, approximately 1200 small farm dams (otherwise known as water-holes) have been built. Oyster farming commenced in the estuary in 1985. We now come to the crucial question. If water in the catchment was to be reallocated (for example, a greater amount impounded to be used by the upstream farmers), what would be the overall (or net) economic result for the river basin community? In the first instance, would there be any discernable effects downstream? Would additional water made available to farmers be worth more in economic terms than any possible adverse effects on the estuary and nearby coastal environments? That is, if it was proved that the oyster growers would lose production and profits as a consequence of a large upstream dam, what would the monetary value of that loss be compared to any increase in value of production from the upland farms? This is a typical economic question that can be answered if sufficient reliable data can be gathered. We will come to explain how we approached this matter in a practical sense. A fundamental issue for the proposed dam was: would it be used for flood or trickle irrigation? The speed at which water flowed through the system would differ accordingly. There were other questions that needed to be addressed before the impacts of changes in water flow could be analysed.

A considerable proportion of the catchment is in near-natural condition due to its status as a Defence Force reserve. Don't imagine soldiers and tanks crashing through the forests. Not once in a period of three years did the research team see any military-type activity as they poked into the back blocks, whenever roads and tracks would take them, and occasionally along imagined tracks. That doesn't necessarily mean there wasn't any military activity; Defence personnel are not required to advertise their presence.

There are other extensive areas of land in a forestry reserve status, as well as large private nature conservation blocks. These near-natural areas have 'non-use' environmental values due to their provision of ecosystem goods and services that the owners value. We were keen to measure these values as they are an important feature of our catchment. In fact, we started with the question of what are the dollar values of ecosystem goods and services in the catchment rather than the conflict in water values. The latter issue became more important as our research progressed. If the ecosystem goods and services could be measured and found to be of high value there would a case to highlight these values in terms of any change in water flows.

In addition to, or in conjunction with, the economic values there are the social values, particularly in the closely settled parts of the catchment where actual communities exist. Could these values be measured as social capital making a contribution (an undervalued one) to the welfare of people living in the catchment? As explained earlier, we can think of social capital as the glue that keeps a community intact – trust, mutual concern, helping neighbours and much more, and the economic 'value' of social capital is that it reduces policing, legal and business costs ('transaction' costs, as the economists call them). Sharing of farm equipment, doing unpaid work on a neighbour's farm, or just sharing ideas and providing friendly advice all have a positive economic impact and build and maintain social capital. Measured in monetary terms or not, it is nice to live in a society with high social capital. The economic benefits are secondary considerations. We were to discover that the dispute over water allocation in the catchment had the potential to do damage to the existing social capital.

The matters we have just mentioned were fascinating issues and some questions remain unanswered even though we have completed our extensive research. We do believe we researched the most important ones, as you will discover. Similar issues to the ones we encountered arise throughout the world where human–environment relationships are poorly understood and not mitigated. In this sense, there is nothing unique about the Little Swanport catchment. If the team, hand-picked for this project, were able to unravel the human–environment relationships in the Little Swanport catchment, why not elsewhere? We hope to illustrate this potential in the following chapters. If we are right in suggesting that our approach has universal application, this is the potential spin-off from our research.

Not top down: nuts and bolts

A final word: how did we get to be involved in this fascinating research? The real or perceived problems in the Little Swanport catchment attracted the attention of two public-interest research funding bodies. These bodies were Land and Water Australia, a research funder that targets natural resource problems, and the Fisheries Research and Development Corporation, with a research interest in seafood production. In the Little Swanport catchment there was a natural resource allocation matter to deal with. Water and land were involved, and there was an important seafood industry in the catchment. A research project was clearly called for, and who better to fund it than these two organisations. The team to develop the research framework was put together. That was done, the researchers were commissioned and the field work undertaken, and now we are reporting the results.

A final word on the field work for the economic and social research we undertook. Our goal, as with any keen researcher wanting data from a relatively small population, was a census, to interview everyone living in the catchment. Researchers, even seasoned ones, tend to remain optimists, always hoping for the 100 per cent coverage. This is in impossible goal that even the best-funded (and powerful) social–economic research bodies cannot achieve.

We were provided with an address (mailing) list of residents in the catchment. This we used for the first stage of a questionnaire-based survey. There were the expected 'return to sender' responses. The properties had been sold between the construction of the mailing list and the time we came to use it. There were a few incorrect addresses – some properties were not in the catchment – and there were, as expected, the non-responses.

Stage two saw the in situ field work. All non-respondent residents were visited over a period of some weeks. Face-to-face interviews were conducted when that was permitted;

otherwise residents were given a replacement questionnaire – if not at home, a questionnaire was left at the residence. Our response rate increased, yet it was still not high enough.

Stage three was a repeat of stage two, this time targeting the remaining non-respondents. This took place some months after the earlier phase. By this perseverance we obtained a 65 per cent response rate overall. It was lowest for the dozen very large farms (at just under 50 per cent), higher for the rural residential group (at 80 per cent), and 60 per cent for the large number of small-to-medium farmers. Because the rural residential respondents are excluded from the following analysis of agricultural activities in the catchment, overall bias is not a serious concern. Finally, to complete our interviews of farmers, a representative sample of businesses in the surrounding towns were subjected to interviews.

9

The river system and water management

Christine Crawford

The Little Swanport River and its tributaries are the backbone of the catchment and their flowing water is the lifeblood. Although we know that water is essential to many activities across the catchment, this is never more obvious than when it is not available, such as during the drought in the middle of the first decade of the twenty-first century.

Although a small catchment with relatively few people compared to some of the intensely farmed and settled areas in northern Tasmania, information available on the Little Swanport River and its tributaries is comparatively abundant. This is largely a reflection of the people living in the catchment; they have been proactive in developing a community-based management plan for the catchment, in addition to the Water Management Plan developed by the Tasmanian Government. As we discussed previously, water has been a divisive issue in this catchment, and as a consequence the Tasmanian Government has conducted a number of studies on water-related matters across the catchment; these are available on the Department of Primary Industries, Parks, Water and Environment website (www.dpiw.tas.gov.au). Of particular note is the comprehensive State of Rivers Report for Little Swanport (DPIW 2006b), which clearly describes the hydrology, ecosystem values, geomorphology, water quality and aquatic ecology of the catchment. Much of the information in this chapter has been summarised from this report and others produced for the holistic environmental flows project which we discuss later (DPIW 2007c).

River ecology: landforms and river flow

We will start with a brief description of the landforms and underlying geology because they are important in determining the condition of the river.

Water as it flows across a catchment also transports sediments and organic matter, and the rate of transport depends on the availability of the different sediment types and the river flow. This in turn determines the shape of channels and floodplains, and the plants and animals they contain. The predominant geology in the catchment is dolerite, and the main source of sediment for the river comes from the erosion of dolerite soils from cleared areas in the upper catchment (DPIW 2007a). The uppermost parts of the catchment, approximately 350 m above sea level, are low relief, rolling hills with dolerite, sandstone and mudstone soils. Broad river plains – marshes, swamps and lagoons – connected by distinct small channels contain fine sediments and are water storage areas. Below this the channels start to connect together on an associated flood plain and the sediments are more mobile. This is a sediment storage and supply region for downstream river reaches and is most susceptible to degradation from changed land use and flow. Much of this upper part of the catchment has been cleared for agriculture and the channels have been modified.

The bottom half of the catchment is steeper, with moderate slope of the river bed and relatively steep valley sides that partly confine the streams, leading to higher energy flows and some flood plain development. The substrate is coarser dolerite cobbles/boulders and the river bed is more diverse, consisting of riffles and pools when flowing. A confined zone above the estuary has steep gorges and no flood plains. The river bed is bedrock with some boulder/bedrock cascades and deep, permanent pools. This landscape generally has poor soils and has not been modified for agriculture (DPIW 2006b; 2007a).

Flows

Good flow data are available for 1971–90 from a gauge installed in the river 3 km upstream from the Tasman Highway bridge and just above the upper limit of seawater penetration. During this period annual flows ranged from 12 000 to 190 000 ML, with an annual mean of 77 000 ML. The mean annual total catchment yield of water at the mouth of the river was estimated to be 113 000 ML (including estimated contributions from tributaries downstream of the gauging site and from several independent streams such as the Ravensdale Rivulet) (DPIWE 2003). Seasonal flows during this period were highly variable but winter (May–October) flows were generally higher than summer (November–April). Periods of no flow did occur, mostly over summer.

This gauge was removed by the Inland Fisheries Service in 1991 because it restricted the movements of migratory fish up and down the river. A new gauge was installed in 2004 near the previous gauge; however, it is yet to be rated for large floods. Annual flows (ML) recorded at this gauge since 2004 are shown in Table 9.1. The effect of the drought conditions in 2006–07 on flows is obvious. In 2004 a second gauge was installed by the Tasmanian Government higher in the catchment, downstream of Eastern Marshes Rivulet.

Using these flow data, Sinclair Knight Merz, consultants employed by DPIW, prepared a water balance model for the catchment as part of the development of the Water Management Plan for Little Swanport. They produced a daily flow series over 100 years for the natural flow regime, and for current and predicted levels of water use (Sinclair Knight Merz 2004). Under natural flow conditions the river system shows a rapid response to rainfall and is typical of 'flashy' rivers with highly variable flows. There were also periods of no flow, and these cease-to-flow periods occurred throughout the catchment. According to Sinclair Knight Merz (2004) under current levels of water use, medium to high flows are similar to those under natural conditions, but lower flows (occurring more than 70 per cent of the time) are different. In particular, the model showed that the river never ceases to flow under current water use, most likely due to runoff from irrigation in the upper parts of the catchment. These low flows can be up to two times higher under today's water use conditions than natural flows, especially during the dry summer months. The flow patterns are also now much less variable in summer than under natural conditions. This change in flows is greatest in the upper catchment and is not dissimilar to other rivers in

Table 9.1. Annual river flow at the bottom gauge just above the estuary in 2004–07

Year	River flow (ML)
2004	31 361
2005	75 258
2006	1238
2007	4258

south-eastern Tasmania, such as the Coal River. However, it has been difficult to ascertain whether the flows and levels of water use predicted by Sinclair Knight Merz (2004) are actually occurring.

An analysis by DPIW of average flows at the two gauging stations in the catchment under natural and current water use has shown that current water use reduces the average volume and speed at which the river rises, but not the time it takes to rise (DPIW 2006b). The Department suggests this occurs because during a precipitation event farm dams collect some of the water from off-stream sources that would otherwise flow to the river. Thus there is a reduction in volume of water reaching the river. However, the variability of flows is similar between natural and present water use conditions.

We will hear more about river flow in Chapter 11, which deals with flows into the estuary.

Water quality

An intensive survey of water quality across the catchment was conducted by DPIW between September 2003 and December 2004 as part of their State of Rivers assessment of the Little Swanport. Twenty-five sites were sampled monthly for electrical conductivity, pH, temperature, dissolved oxygen and nutrients. Snapshot surveys of the entire catchment were undertaken in summer and winter during periods of stable flow, measuring nutrients, metals (e.g. cadmium, copper, iron) and bacterial concentrations. The ionic composition of the water was also sampled quarterly from a subset of 11 sites (DPIW 2006b).

The results of this water quality assessment are described in detail in the State of Rivers Report for the upper, middle and lower catchment (DPIW 2006b). In the upper catchment water quality is strongly influenced by ephemeral flows, that is intermittently flowing streams and channels, and by the land use, especially agriculture and associated removal of riparian vegetation. Overall, water quality in this area is the poorest of the catchment. The low to no flows in conjunction with loss of riparian vegetation resulted in high temperatures, large variations in dissolved oxygen, and high turbidity and nutrient levels after significant rainfall. Extremely high nutrient concentrations also periodically occurred as a result of local influences, in particular from livestock accessing waterways.

Of particular concern is the high salinity content of the water, with median values exceeding 1400 μS/cm at all sites (salinity >700 μS/cm is unsuitable for horticulture and >1500 μS/cm has direct adverse biological effects) (DPIW 2006b). This confirmed previous sporadic monitoring in the catchment which identified high conductivity (saline) concentrations at most sites in the catchment, especially in the upper Little Swanport River and upper Eastern Marshes Rivulet catchments. In contrast to many other catchments, conductivity increased towards the top of the catchment, and appears to be strongly related to groundwater. Extensive land clearing in the upper catchment has probably contributed to groundwater salt inputs. Conductivity was highest during low flow periods when the influences of saline groundwater were greatest and significantly lower during and after rainfall which increased dilution.

Water quality generally improved in the middle catchment, largely due to changes in geomorphology, increased base flow and increasing occurrence of riparian vegetation, although the poor quality of water from Nutting Garden Rivulet did have an influence. In the lower catchment the rugged and steep terrain is not suitable for agriculture and much of the native riparian vegetation remains. As a consequence of this and increased inflow, the water quality was considerably improved, with a continual reduction in salinity and in median and maximum concentrations of nitrogen.

A statewide Baseline Water Quality Monitoring Program (BWQMP) was established by DPIW in 2003 to monitor a range of water quality parameters at 52 sites around Tasmania, and additional automatic flood sampling at 37 of these sites. This baseline monitoring program did not include estuaries. Little Swanport River was monitored downstream of Eastern Marshes Rivulet (below all major agricultural and irrigation areas) and 150 m downstream of the weir just above the estuary. In-stream sensors continuously monitored water temperature, conductivity (salinity) and turbidity at both sites, and dissolved oxygen in the middle catchment only. Monthly spot measurements were also taken at these two sites for the above variables and for pH, and water samples were also collected for later nutrient analyses (nitrogen and phosphorus). Automatic flow-triggered flood samplers have been installed at these two gauging sites; however they have not always functioned during flood events.

DPIW have used these data on ambient water quality collected over three years to develop guideline or trigger values for ambient water quality based on the guidelines (Table 9.2) (ANZECC 2000). These trigger values are calculated using the 80th percentile, whereby 80 per cent of all measurements over the last three years are equal to or less than this value. For example, the trigger value for nitrate at the lower site is 0.051 mg/L, so 80 per cent of all readings over the last three years have been equal to or lower than this value. Hence any reading above this should trigger further investigation. Environmental variables such as dissolved oxygen and pH have both upper and lower values in relation to the quality of the water, and thus have both 80th and 20th percentile values. However, it should be recognised that these trigger values have limitations; for example they do not take into account whether these physical and chemical parameters are having any impact on the ecology of the system. Also they do not show whether there have been changes in water quality over time and thus do not show whether these trigger values are being set for a relatively pristine site or whether it is in fact already degraded.

Herbicides and pesticides, including those most frequently used in agriculture and forestry activities in Tasmania, have been monitored quarterly since 2005 by DPIW in a number of catchments around Tasmania, including Little Swanport. Automatic flood samplers are also located in this catchment. No herbicides or pesticides have been detected in samples collected quarterly from July 2005 in Little Swanport. Flood samples collected in June 2005 detected Terbacil, a selective, systemic herbicide used to control annual grasses, broad-leaved weeds and some perennial weeds in established orchards and eucalypt forest plantations (DPIW nd), at concentrations much lower than the guideline trigger value.

Ground water

There is virtually no information (certainly of a reliable nature) on groundwater quality or quantity in the catchment and this water supply has not been included in the determination of environmental water requirements or the Water Management Plan. However, DPIW has committed to the establishment of a groundwater use register, largely wells and bores and quantity and rates of abstraction. Leaman (2004) has been particularly critical of the lack of recognition of the importance of groundwater in the Water Management Plan.

Aquatic ecology

Along with measures of water quality, the ecology of aquatic systems was investigated to show what effects physical and chemical changes in water quality are having on the biological health of the system. The two communities that are commonly used to assess river health are aquatic macroinvertebrate fauna and fish fauna. A standard protocol used across

Table 9.2. Trigger values for ambient water quality at the two gauging sites on the Little Swanport River

	Trigger values for freshwater	
Parameter	**20th percentile**	**80th percentile**
Site 2235: Little Swanport River 150 m Downstream Weir		
Water temperature °C	8	20
Turbidity NTU	1	6
Field Cond at TRef 25 µS/cm (25 TRef)	484	799
pH field – sensor TC	7.5	8.2
Dissolved oxygen mg/L	9	12
Dissolved oxygen percent saturation	91	106
Total nitrogen as N mg/L	0.368	0.647
Total phosphorus mg/L	0.006	0.016
Phosphorus, dissolved reactive as P mg/L	0.003	0.005
Nitrate as N mg/L	0.004	0.051
Nitrite as N mg/L	<0.002	<0.002
Ammonia as N mg/L	0.006	0.013
Site 2212: Little Swanport River Downstream Eastern Marshes Rivulet		
Water temperature °C	7	17
Turbidity NTU	1	6
Field Cond at TRef 25 µS/cm (25 TRef)	714	1064
pH field – sensor TC	7.4	7.9
Dissolved oxygen mg/L	8	11
Dissolved oxygen percent saturation	84	101
Total nitrogen as N mg/L	0.509	0.960
Total phosphorus mg/L	0.008	0.018
Phosphorus, dissolved reactive as P mg/L	0.002	0.005
Nitrate as N mg/L	<0.002	0.029
Nitrite as N mg/L	<0.002	<0.002
Ammonia as N mg/L	0.006	0.011

Australia for assessing river health, the AusRivAS (Australian River Assessment System) model, has been used in the catchment. This method is based on comparing the macroinvertebrate community at sites in the river with those from relatively pristine reference rivers nearby. Macroinvertebrates are particularly useful for monitoring aquatic health because their largely sedentary nature and relatively long life cycles show an integrated response to a variety of environmental variables over time. Also the response of many macroinvertebrates to different environmental conditions is relatively well known.

In 1998 the macroinvertberate fauna in riffle and edgewater habitat at eight sites across the catchment were investigated. Five sites were classified as equivalent to reference sites (i.e. good condition), while the Ravensdale Rivulet at Tasman Highway and Crichton Creek at Swanston Road were assessed as being severely impacted, and Eastern Marshes Rivulet was significantly impaired compared with reference sites, probably because of the ephemeral nature of water flow. In spring 2003 and autumn 2004, 25 sites surveyed across the catchment showed considerable variation in condition. There was a general trend of lower diversity and reduced abundance of pollution-intolerant species in the upper catchment compared with the lower and middle reaches, and this generally corresponded with greater levels of habitat modification and degradation and poorer water quality. However, because of the highly variable natural flows, including periodic complete cessation of flow, it is difficult to separate naturally occurring flow disturbances on macroinvertebrate fauna from human-induced impacts.

Continued monitoring of river health using AUSRIVAS macroinvertebrate faunal methods in the middle reaches of the catchment in 2004–06 showed significantly impaired riffle habitat and a slightly better edge water habitat, both of which were probably exacerbated by low flows, which were so low in 2007 that assessment was not feasible.[1]

The river was surveyed for fish in 2001 (Pinto 2001) and again in 2003–04 as part of the State of Rivers report. Three introduced species, brown trout, redfin perch and tench, were caught. Native fish species included pouched lamprey, short-finned and long-finned eels, sandy or freshwater flathead, and the galaxiids jollytails and spotted galaxias. The endangered Australian grayling has been reported from the river, and another species also listed as endangered, Swan galaxias, has been introduced into tributaries of the upper Little Swanport River where the predatory brown trout and redfin perch do not occur. All these native species, except for the Swan galaxias, are migratory and move between the river and the sea to complete their life cycle. There was a general pattern of decreasing diversity with increasing distance upstream, which may relate to the better condition of the river habitat in the middle and lower catchment, including good in-stream cover and riparian vegetation, a greater range of substrate and habitat types, and closer proximity to the sea. Exotic species were more common in the generally more degraded upstream habitat with little riparian vegetation, low habitat diversity, and silt substrates (DPIW 2006b). Platypus also occur throughout the catchment.

As part of the holistic environmental flows project (described later), a review was conducted on the flow and flow-related habitat requirements of Tasmanian native and introduced freshwater fish, including those at Little Swanport (DPIW 2007c). This review concluded that to maintain the fish populations in the Little Swanport catchment, rising flows and flood events are required, as well as maintenance of base flows in all seasons. For example, floods are necessary in autumn–winter to trigger spawning behaviour and provide access to spawning habitat, to hatch eggs and to disperse larvae. They are also required in winter–early summer to initiate upstream movement of several galaxiid species, to trigger upstream spawning migrations of whitebait species *Lovettia seallii*, downstream migration of adult eels, and upstream movements of elvers and juveniles. Flood events are also required to maintain the habitat that produces benthic invertebrate food for fish.

Riparian vegetation

Riparian vegetation alongside the river, its tributaries and the estuary is important to the condition of the waterway because it is a buffer between the water and the dominant land activities. This vegetation helps stabilise the river banks and filters out nutrients and sediments before they reach the water. The riparian vegetation along the Little Swanport River is diverse, with many endemic species in most sections of the catchment, but especially in the lower middle reaches where there is little land development. However, in agricultural areas of the upper catchment riparian vegetation is scarce and generally contains weeds, such as gorse and blackberry (DPIWE 2003). A survey of riparian vegetation at three sites along the river found that this vegetation reflected the surrounding land use – from close to natural at sites with little or no evidence of impact, through semi-natural, rough grazing with evidence of land use, to largely cleared for grazing (DPIW 2007c). However, the in-stream plant communities varied little between the three sites along the river even though flow intensities were of different magnitude and channel characteristics differed.

Water management: protected environmental values

Protected Environmental Values (PEVs) for surface waters have been developed by the Tasmanian government in conjunction with the Little Swanport community for the different land uses and values of the waterway in the catchment (DPIWE 2005b). They are a requirement of the *State Policy on Water Quality Management 1997* and as part of the water management planning process to develop Water Quality Objectives and associated targets. PEVs provide a strategic framework for water quality management and focus the attention of industry, government and community on the long-term sustainable use of surface waters. However, the PEVs tend to be general and differ little between catchments around Tasmania. The PEVs for Little Swanport are, at a minimum, that the water quality shall be managed to provide water of a physical and chemical nature to support a healthy but modified aquatic ecosystem from which edible fish are harvested; and which will allow people to safely engage in secondary contact recreation activities such as paddling or fishing in aesthetically pleasing waters.

Water values recognised by the community as part of developing the Water Management Plan include as highest priorities maintaining native species/populations and their habitats, especially the black bream recreational fishery, protecting wetlands in the upper catchment and saltmarsh in the estuary, retaining native vegetation, protecting existing water rights, and maintaining clean flowing water, free of chemicals and waste products.

Additional community objectives for water quality and quantity were developed for the Little Swanport Community Catchment Management Plan, a separate process from the water management planning. However, because both plans were developed at much the same time, many members of the community tended to combine the two processes together or were simply confused as to what was going on.

Water management plan

Because of the issues around the use of water in Little Swanport, this catchment was one of the first in Tasmania to have a Water Management Plan developed. It was prepared by DPIWE to provide a framework for managing the catchment's water resources in accordance with the *Water Management Act 1999* and the *State Policy on Water Quality Management 1997*. Interestingly, and largely for historical reasons, water quality and quantity in Tasmania are managed under two separate legislative frameworks, the *Water Management Act 1999* and the *Environmental Management and Pollution Control Act 1994* respectively, even though they are integrally related. As a consequence, the development of Environmental Water Rights (which underpin the plan and are described in more detail later) does not include any assessment of water quality (Pinto 2001). The Water Management Plan is a statutory plan affecting everyone using water in or from the catchment. It was released in draft form for public comment in September 2004. The final plan was released by the Minister for Primary Industries and Water in June 2006 and is scheduled for review in five years' time (DPIW 2006a).

The specific objectives of the Plan are:

- Environmental – maintain natural refuges for in-stream biotic communities during naturally dry periods and protect flow regimes important for geomorphic processes, riparian vegetation, flood plain processes and estuarine processes dependent on freshwater inputs
- Water usage and development – protect water resources for current and future stock and domestic use, provide water to meet existing and future licensed use, establish water availability for potential sustainable economic development, and ensure that future allocations do not significantly impact on the quality of water
- Management of the water resource – improve knowledge of the licensed water use and of the catchment's water resources and aquatic environment.

The Plan makes water management provisions for surface water licenses, allocations, flood harvesting and restrictions on taking water during specified low flow conditions according to set priorities for water use. The priorities range from Surety 1 (highest) – for water essential for town water supplies, domestic purposes, consumption by livestock and firefighting; Surety 2 – water required to sustain ecosystems dependent on water resources; to Surety 5 – rights to take water for irrigation and other commercial purposes, by direct abstraction or for winter storage in dams; and Surety 6 – harvested during flood events.

Based on a daily water balance model developed for the catchment (Sinclair Knight Merz 2004), a flow regime that meets the objectives of the Plan was prepared with the following provisions:

1. A total catchment allocation at Surety 5 of 6084 ML/year
2. The adoption of summer (recorded flow ≤ 7.6 ML/day for November – April) and winter (recorded flow ≤ 9.5 ML/day for May–October) cease-to-takes on all Surety 5 allocations
3. A flood harvesting rule that allows up to 10 per cent of the total flood volume (ML/day) to be available when flow exceeds 2000 ML/day.

The Plan includes guidelines for allocating water to major sub-catchments to ensure that the long-term average flow from these sub-catchments remains above low-risk environmental flows, although the total amount cannot exceed 6084 ML/year.

The allocation of water, however, is confusing because 3158 ML of water allocated per year actually comes from the Hobbs Lagoon system in the neighbouring Prosser catchment and is applied in the Nutting Garden Rivulet catchment, and any water not taken up eventually flows down the Little Swanport River. This allocation is from the Prosser catchment on the DPIW database but is managed via the Water Management Plan in the Little Swanport catchment.

The Plan allows for a 57 per cent increase of current Surety 5 allocation, which would result in a 2.3 per cent reduction of the mean annual discharge over current estimated condition in 2003 to the Little Swanport estuary. This represents a 6.2 per cent reduction in the mean annual discharge from the modelled natural condition. The government considers this water regime provides adequate low flows during naturally dry periods to maintain in-stream habitat, including refuges for in-stream animal and plant communities, because low flows are protected according to Provision 2 (no water can be extracted for irrigation when flows get below the prescribed limit). According to the Water Management Plan, modelling indicates that water use under the Plan has a negligible long-term effect on the current flow regime, including frequency of floods on an annual or longer time scale. In dry years the modelling shows a slight reduction in the magnitude of peak discharges under the Plan, but no effect on the timing, frequency, duration or predictability of flows (DPIW 2006a).

According to the Waterways Monitoring Report for 2007, the total water allocation in 2007 was 3494 ML, with the bulk of the water (3230 ML) going to irrigation and 264 ML to stock and domestic supplies. Only 30 ML was extracted directly from rivers and streams; the rest was held within constructed storages. However, as water gauges have not been installed in most regions of the catchment, even though it was a requirement of the Water Management Plan for this to happen within 12 months of the release of the plan, the actual take and storage of water is unknown. The monthly discharge of water in 2007 was significantly lower than the historical average for most months, but particularly over winter and spring.

As to be expected with any statutory plans for the management of a scare resource, there are opponents to the Plan. The main criticisms of the Plan are based around the data used, or lack thereof, and underlying assumptions of the water balance model (Leaman 2004). First, the model does not take into account groundwater flows as the extent of these is not known. Another criticism is that the Plan is based on flow data from July 1971 to January 1990 and extended by rainfall runoff modelling to the period 1 January 1900 to 31 December 2001. According to Leaman (2004) this gauging period was a time of transition and atypical, and the rainfall since 1990 has been significantly less than in the years prior. He compared water yield patterns in the Little Swanport with adjacent catchments where there are continuous flow data and concluded that water yields had decreased by up to 50 per cent since 1990. Changes in the timing of rainfall events are also considered to be important, as loss of autumn rainfall means groundwater is not topped up after summer.

The major concern, particularly in relation to implementing the Plan, is that there are still no records of water use across the catchment. Although the Plan stipulates that licensees who extract water from any watercourse in the catchment for irrigation or any other commercial purpose are required to install an appropriate meter within 12 months to measure the volume of water extracted, this has not happened. Without records of the amount of water being extracted, it is impossible to know whether the Plan is being adhered to.

Environmental water requirements

Underpinning the Water Management Plan is the assessment of Environmental Water Requirements (EWRs). These are defined in the Water Management Plan as:

> *descriptions of the water regimes needed to sustain the ecological values of aquatic ecosystems at a low level of risk. These descriptions are developed through the application of scientific methods and techniques or through the application of local knowledge based on many years of observations. (DPIW 2006a, p. 4)*

The ecological assessment of low flow requirements for the Little Swanport River was based on both community and state technical ecological values, which included maintenance of habitat for the common jollytail (*Galaxias maculatus*), short-finned and long-finned eel, platypus and macroinvertebrates in the river. Recreational values specifically targeted were to maintain the habitat for brown trout populations (a species introduced into Tasmania in the 1800s). The EWRs were determined from modelling the amount of wetted useable areas available for in-stream aquatic organisms of high ecosystem value and determining the minimum flow required to maintain these in-stream populations at specified levels of risk.

Although other fish species were considered to be of ecological value, they were not included in the risk assessment because of lack of data on their specific habitat requirements.

These include the spotted and Swan galaxiids, *G. fontanus* and *G. truttaceus*. Flow requirements in estuaries, in particular, are not included because virtually nothing is known about the estuarine habitat requirements of several freshwater species that spawn in estuaries or migrate through to the sea as either larvae or juveniles. These include *G. truttaceus*, the common jollytail, whitebait *Lovettia sealii*, freshwater flathead (also known as sandy or congoli), Australian grayling, and two species of lampreys. However, it is known that adults of short-finned and long-finned eels mature and migrate downstream to the sea during the autumn when floods are more common. It is therefore important that the autumn and winter flooding flows are allowed to flow through the estuary.

Holistic flows framework

For many years low flow periods have been the priority area for management in waterways around Australia, as described above for EWRs. However, it has been increasingly recognised that medium and high flows are important to healthy functioning ecosystems. As a consequence DPIW has developed a framework to guide the assessment and recommendation of holistic environmental flows regime for Tasmanian catchments. The Tasmanian Environmental Flows Framework (TEFF) (DPIW 2007c) builds on the EWR assessment for minimum flows. It also follows the natural flow paradigm of maintaining the pattern of the natural flow as far as possible in order to preserve freshwater-dependent values of a system.

While large floods will flow through a system largely untouched (except in catchments with very large water storage systems), the magnitude, frequency and duration of smaller flow events can be affected by water abstraction and collection. The TEFF attempts to link the physical and biological characteristics of a river to specific flows. Flows required to maintain riparian vegetation, geomorphological characteristics, in-stream habitats, spawning and migratory requirements of native freshwater fish species and the freshwater requirements for estuaries were assessed. Three river systems were targeted for field assessments and hydrological analyses to develop holistic environmental flow regimes, one of which was the Little Swanport. This framework is being further developed and tested in a follow-on project, the Tasmanian Environmental Flows (TEFlows) Project, which aims to determine how variation in small-scale flows might drive the physical structure and biological function of a river. This work is in progress, with the Little Swanport River continuing to be one of the trial catchments.

10

The Little Swanport estuary

Christine Crawford, John Gibson and Barry Gallagher

A photograph from the mouth looking towards the river entrance provides an overview of the topography of the Little Swanport estuary (Figure 10.1). A lack of urban development but intensive clearing to the water's edge is obvious, as is the oyster racking between the entrance bar and Ram Island. The river entrance, north of the Duck Islands, is very small compared with the broad intertidal shallows of the estuary. The area shown in Figure 10.1 is commonly referred to as the Little Swanport estuary, although saline waters regularly penetrate several km up Little Swanport River. Because of the lack of development, other than agriculture and aquaculture, this estuary is considered by many to be picturesque and serene.

Changes in the estuary over the last 7000 years

All estuaries are in a constant state of change. Little Swanport as we see it today is the end result of long-term (millions of years: landform formation, erosion) and shorter term (tens of thousands of years: glacial cycles, rising and lowering sea level) processes. Further change has occurred since sea level stabilised approximately 7000 years ago. Understanding these changes and how they have occurred will allow us to understand and predict modern and future change with greater certainty.

Aboriginal settlement of Tasmania began around 35 000 years ago, and thus there has been human influence in the area for the entire existence of the estuary as it is today. The earliest evidence of human habitation around Little Swanport has been inferred from the mounds of the native oyster, *Ostrea angasi*, just inside the estuary's mouth. The age of the midden has been estimated to be approximately 4500 years (Lourandos 1970), but habitation in the area has probably occurred for much longer.

Aboriginal fire practices have been shown elsewhere in Tasmania to increase landscape erosion and therefore sediment deposition in estuaries (McIntosh *et al.* 2004). As a considerable proportion of the Tasmanian Aboriginal population lived in the little Swanport area around the mid-nineteenth century (estimated population 500, Taylor 1892), there was probably a similar impact in this catchment. More direct evidence of extensive pre-European burning comes from studies of the carbon to nitrogen (C:N) ratios of soils from dry eucalypt forests in north-eastern Tasmania. The high C:N ratios are consistent with much earlier Tasmanian Aboriginal settlement and fire practices in Tasmania than the more recent arrival of the Maori to New Zealand (Macintosh *et al.* 2005). Thus, it is likely that

Figure 10.1 Map of Little Swanport estuary. Photograph by Francisco Neira.

Little Swanport has experienced an increased rate of sediment deposition over that of entirely natural conditions since sea level stabilised approximately 7000 years ago.

We might therefore expect that the start of European colonisation within the area in the 1830s (von Platen 2008) would have led to a reduction in fire frequency and therefore a concomitant decrease in the rate of land erosion. Indeed, evidence from tree rings in the Little Swanport catchment indicates a significant fall in fire frequency (von Platen 2004). However, early colonisation also brought with it a frenzy of land clearance for farming from the 1830s to the 1850s. This is likely to have led to an increase in soil erosion, particularly in the upper catchment where most of the clearing occurred, resulting in the transport of eroded sediment to the lower estuary.

Estuary sedimentology

So what are the consequences of this increased erosion? In general, the distribution of sediment within an estuary is determined by the relative strengths of river flow, wave action and tidal currents. Lewis (2006) showed for Pittwater estuary (south-east Tasmania) that land clearance led to an encroachment in the estuary of mud derived from the catchment, as distinct from sand from the sea. In Little Swanport, where the narrow river channel abruptly enters the wider expanse of the main estuarine water body the flow velocity reduces suddenly, allowing the sediment and fine particles transported down river to settle out in this less turbulent environment. This process led to the formation of the Duck Islands in the region of high deposition.

Any subsequent redistribution of sediment within the estuary is determined by wave action, tidal currents, wind, and the depth and stability of the sediments. As a consequence, we would expect the terrestrially derived mud deposited in the main body of the upper estuary to be relatively stable, and only catastrophic flood events or tidal surges due to storms or possibly tsunamis could provide sufficient energy for any redistribution (Switzer and Jones 2008). Seagrass and macrophytic algae also stabilise sediments, partly reducing the level of sediment resuspension. They can also promote an increase in the rate of gross sedimentation by reducing the local current velocity within the seagrass meadows, resulting in the settlement of finer particles (Garcia *et al.* 2003).

Catastrophic changes over the last 100 years

The evolution of Little Swanport, like that of many other estuaries, appears to be dominated by rapid, event-based changes overlain on gradual, continuous change. This is evident in sediment cores taken from within the estuary. These cores suggest that catastrophic events have both eroded and deposited sediments within the upper portion of the estuary and have affected the depth and width of the bar over the last century. They indicate that the two major flood events of the last century, in 1923 and 1960, not only added new sediment to the estuary but also eroded and redeposited terrestrially derived sediment, along with old (550–750 years) intertidal cockle shells. These floods led to a deposition of terrestrially derived material closer to the entrance of the estuary, which was inferred from the sustained transitions from sand to mud-dominated sediment in the cores corresponding to the times of the floods. The sedimentation rate in the deep basin is estimated from the sediment cores to be approximately 0.9 cm per year. Based on this sedimentation rate, it would take in the order of 1000 years for the estuary to largely fill with sediment.

The estuary today

Some statistics for the Little Swanport estuary include a water area of 4.3 km^2, intertidal area of 1.7 km^2, 31.3 km perimeter, and maximum depth of 8 m. The total volume of the estuary is estimated to be 7 900 000 m^3 and maximum tidal height is about 1.2 m, with the high tide approximately 60 minutes behind the permanent measuring site at Spring Bay. The tidal prism, which is the amount of water that enters and leaves an estuary through one tidal cycle, is approximately 3 400 000 m^3, or 43 per cent of the volume at high tide. The main freshwater input into Little Swanport is the Little Swanport River, though other, smaller creeks enter the estuary directly. The area of the drainage basin of the estuary is 723 km^2 of which 105 km^2, or approximately 15 per cent of the total, is direct drainage.

The Little Swanport River approaches the estuary in a well defined channel running through rocky hills. The upper reaches of the estuary are mostly shallow (<3 m) with a few deep holes, and largely cobble bottom covered in fine silt. Noticeable halo- and thermoclines are often located close to the surface (0.3 m) in times of low freshwater flow and extending deeper (to 1.5 m) following high rainfall in the catchment. Above the highway road bridge the river broadens out on to a small plain, while under and just seaward of the bridge is a very shallow rock and cobble riffle bed which can impede the movement of fish to and from the estuary during low flows. A complex fluvial delta has built up over the thousands of years of the estuary's existence, resulting in the formation of at least four major islands separated by shallow channels – the Duck Islands – of total area 31.4 ha. While most water flow at present occurs along the southern side of the estuary, in previous times the positions of the islands and major flow were probably different, as evidenced by the relatively deep channels between the islands and the deep, linear feature near the shore of the estuary to the north of the Duck Islands.

Close to the point at which the river enters the main estuary is shallow Watch House Bay. This bay is highly productive, harbouring extensive stands of seagrass, *Ruppia* sp. Sedimentation of both inorganic (sand, silt and clay) and organic (dead seagrass) material is slowly filling up this bay, and it will eventually become largely salt marsh unless rising sea level keeps pace with the rate of infilling. Large areas of salt marshes border the shallow mud flats and Duck Islands in this fluvial delta region.

Seaward of the fluvial delta the estuary is deeper (>2 m at low tide), and the bottom is composed of fine mud. A trench containing the deepest water of the estuary, at around 8 m, extends to the south of Ram Island. This trench is probably the old river valley that has progressively infilled since the formation of the estuary. The middle estuary is predominately a shallow lagoon with dense seagrass, *Zostera tasmanica*, beds that are mostly inaccessible at low tide. Ram Island is the largest island in the estuary (27.6 ha), and is pre-existing high land rather than a depositional feature.

Towards the mouth of the estuary the depth reduces as the tidal delta is reached. The sediment changes from fine mud to sand. As with the fluvial delta, there are extensive areas that are exposed at low tide, but in this case they are typically covered by water at high tide. The disposition of these areas does not appear to have changed much over the last 50 years, but aerial photos indicate that the entrance was wider 50 years ago and has progressively narrowed.

The lower estuary comprises a mosaic of habitats which include large expanses of *Zostera* seagrass, coarse sand, shell beds, and fine silt sediments. It is marine dominated and does not experience substantial drops in salinity except during large floods.

The underwater habitats of the main estuary have been mapped by the Tasmanian Aquaculture and Fisheries Institute in 2005 as part of the SeaMap Tasmania program (Mount *et al.* 2005). Bottom sediments and surface dwelling biota of the estuary were mapped using aerial photographs, echo sounding of the seabed and ground truthing with underwater video. The sediments in the lower estuary are predominately sand, with less than 15 per cent silts and clays (i.e. <0.063 mm particle size), whereas the middle estuary consisted of finer sediments with much of this region containing 30–45 per cent silts and clays. The area of each type of habitat is listed in Table 10.1.

Table 10.1. Area of habitat in Little Swanport (from Mount *et al.* 2005)

Little Swanport habitat types	Area (ha)
Low profile reef	1.5
Cobble	6.0
Sand	87.4
Silty sand	52.9
Silt	54.5
Dense seagrass	209.9
Patchy seagrass	76.2
Sparse seagrass	57.1
Vegetated	48.2
Land	14.8
Total	**608.6**

The estuary was classified by Edgar *et al.* (1999) as being of moderate conservation significance but significantly degraded by human activity, based on an assessment of water quality, macroinvertebrate infauna, human population numbers and catchment activities. They concluded that Little Swanport should be available for a range of recreational and commercial activities. Similarly, on the Oz Estuaries database for estuaries around Australia, the condition of the Little Swanport estuary is rated as 'modified' and under moderate to high pressure (http://ozcoasts.gov.au/search_data/detail_result.jsp).

Physical and chemical characteristics

The water quality in the estuary has been monitored on several occasions, including monthly from November 1991 to May 1993 as part of a study assessing the carrying capacity of the estuary for oyster production (Crawford and Mitchell 1999), bimonthly in 1999–2000 (Murphy *et al.* 2003) and approximately monthly since 2004 (Crawford *et al.* 2006). It is also monitored for bacteria on a regular basis as part of the Tasmanian Shellfish Quality Assurance Program (TASQAP), which monitors the water quality of farmed oyster growing areas to ensure that the oysters are safe for human consumption.

During normal river flow the main estuary is largely marine. Temperatures fluctuate according to the seasons, with the greatest range in values occurring in the upper estuary from 23°C in summer to 6°C in winter. Under drought conditions shallow bays, such as Waterhouse Bay, can become hypersaline in summer due to evaporation, and salt water can be detected several km above the highway bridge. Salinities up to 38 have been recorded in the main basin of the estuary. The limited freshwater input during droughts means that there is net importation into the estuary of salt and associated nutrients and biota from the ocean. During significant floods, however, the entire estuary is flushed out and becomes fresh to the estuarine mouth. The estuary generally returns to normal conditions within one week of flooding (Crawford *et al.* 2006).

Dissolved oxygen (DO) concentrations recorded in surface waters from January 2004 have been constantly high; however, in bottom waters they have showed greater variation, especially at sites in the upper estuary and after flood events. During floods in February and July 2004, DO levels were disturbingly low in upper estuary bottom waters but returned to normal within a few days of the flood ceasing. Turbidity levels in the estuary have been low except during flood events, when they reached relatively high levels, especially in the upper estuary. Minor floods, such as annual flushes, clearly change the water quality at the upper estuary sites but this effect declines towards the mouth of the estuary where minimal impact has been detected in terms of salinity, DO or turbidity. However, major floods, such as those in January 2004 and October 2005, affected the water quality of the entire estuary.

Nutrient concentrations (nitrates, phosphates and ammonium) have generally been low throughout the estuary under normal flow conditions. During the drought of the later 2000s (and hence minimal input of nutrients from upstream activities) a seasonal influx of nitrates and phosphates into the estuary from oceanic waters was apparent. Nutrient rich Southern Ocean waters often penetrate up the east coast of Tasmania during winter and spill over into estuaries, whereas the nutrient poor East Australian Current dominates over summer. During flood events the input of nutrients into the estuary can be an order of magnitude higher than normal. For example, under normal conditions the concentration of nitrate in the estuary in 2004 was in the range 0–10 μg/L, except after the flood in January 2004 when the concentration exceeded 250 μg/L. However, almost all the sampling for nutrients in the estuary has been conducted on a monthly basis and the results therefore depend on when the sample was taken in relation to the timing of the flood; peak values are rarely recorded. The concentration of silica in estuarine waters has also been monitored and clearly showed the highest concentrations in the upper estuary, especially during and after floods. This is similar to other estuaries as silica is primarily sourced from the breakdown of ancient rocks and soils. It is essential to the growth of the cell walls of diatoms, microscopic algae that grow in the estuary and are important to the estuarine food web, including the diet of oysters.

In March 2004 the NSW Department of Environment and Conservation conducted a survey of nutrient fluxes between the sediment and overlying water in the Little Swanport estuary (Potts 2004). They found that at this time of year the water column concentrations of inorganic nitrogen (NH_4 and NO*x*) and phosphorus (PO_4) were low. These sediment–water fluxes of inorganic nutrients were lower than in estuaries in New South Wales and were considered to be characteristic of Tasmanian estuaries where the uptake of inorganic nutrients by primary producers (plants) is tightly coupled with their release via breakdown of organic matter on the bottom. In Little Swanport the extensive seagrass beds most likely play a major role in modulating nutrient fluxes between the sediment and water column.

Estuarine ecology

Several indicators of ecological health are commonly used around the world to assess the condition of estuaries. These include chlorophyll a, which is a measure of the biomass of microscopic plant material in estuarine waters. In heavily polluted estuaries receiving bulk loads of nutrients from land-based activities, chlorophyll a values can be in excess of 100 μg/L and large algal blooms are evident. However, in Little Swanport they were generally low and rarely exceeded 3 μg/L. Highest values were recorded in summer and autumn, and lowest in winter. A preliminary analysis of the species of microalgae in the estuary showed that diatoms were dominant over dinoflagellates, indicating that the estuary is relatively healthy and not suffering from eutrophication. The diatom species also suggested that the sediment is regularly suspended in the estuary (Crawford *et al.* 2006).

Macroinvertebrate fauna in the sediments are increasingly being used in Tasmania as an indicator of estuarine condition because soft sediments are the dominant habitat type in estuaries and the macroinvertebrate fauna is relatively stationary and long lived. This fauna in the Little Swanport estuary was briefly examined in 2007–08 when baseline

Table 10.2. Ten most abundant species/taxa of fishes caught in Little Swanport Estuary between March 2004 and June 2005, based on raw unadjusted numbers (from Crawford *et al.* 2006).

	Upstream direction ➡				
Species/taxa	**Near entrance**	**Middle lower**	**Middle upper**	**Above bridge**	**Total**
Acanthopagrus butcheri (black bream)	55	22	114	2243	**2434**
Atherinidae (hardy heads/ prettyfish/silverfish)	516	34	684	771	**2005**
Gobiidae (several species)	213	31	38	254	**536**
Nesogobius spp. (several species)	487		11	8	**506**
Aldrichetta forsteri (yellow-eye mullet)	343	42	57	35	**477**
Galaxias maculates (common jollytail/native trout)	7	6		336	**349**
Rhombosolea tapirina (greenback flounder)	319		4		**323**
Gymnapistes marmoratus (soldier fish/cobbler)	28	7	255		**290**
Afurcogobius tamarensis	41	29	72	123	**265**
Ammotretis rostratus (long snouted flounder/sole)	228	2			**230**
Other species	431	264	274	319	**1288**
Total fish caught	**2668**	**437**	**1509**	**4089**	**8703**
Total species/taxa	**33**	**26**	**25**	**19**	**42**

environmental data were collected from several estuaries in southeastern Tasmania (Temby and Crawford 2008). A total of 26 species were recorded, which was similar to the total for several other estuaries in the region. The average number of species found in each grab decreased with distance up the estuary as the environment changed from predominately marine to fluctuating estuarine conditions and the abundance of invertebrates per grab was relatively low.

Another indicator of estuarine condition that is commonly used is the number of fish species found in an estuary and the abundance of each species. In 2004–05 the fish community of the estuary was sampled every month for 16 months by the Tasmanian Aquaculture and Fisheries Institute. A total of 42 species were caught using a combination of gears: beach seine in shallow water, fyke nets set overnight, and multipanel gill nets of differing mesh size. All fish caught were identified, a proportion measured, and then they were returned to the estuary alive, with only a few of the smaller fish not surviving. The most common species caught were black bream (*Acanthopagrus butcheri*), hardy heads (Atherinidae) and gobies (Gobiidae), which together accounted for 71 per cent of the total catch (Table 10.2). The vast majority of these fish move between the estuarine and marine environments, and only the most abundant species, black bream, spends its entire life cycle in the estuary. This fish, which is the main recreational species in the estuary, migrates to the upper tidal limit to spawn and then the juveniles move back down the estuary.

Fish larvae were also sampled in the estuary in the entrance channel and the upper estuary. A total of 23 species were caught in the plankton nets. Gobiid larvae (possibly from three species) dominated the catch, followed by larvae of black bream and greenback flounder. Larvae of only one freshwater migrant, the freshwater flathead (congoli), were found in the estuary. As sampling was only conducted once a month it is likely that larvae of several other freshwater species moved through the estuary but were missed in the monthly samplings.

The estuary is an important location for waterbirds, including native species and migratory waders. Each year thousands of black swans move to the area over summer, where they breed on the shoreline and feed in the seagrass beds of the upper estuary. Their large numbers mean that swans play a significant role in nutrient cycling in the estuary. Other species commonly found in the estuary include native ducks, white-faced herons, oyster catchers, silver and Pacific gulls, and great and little pied cormorants. The migratory fairy terns also breed in the estuary.

Native oysters

The native, flat oyster, *Ostrea angasi*, has clearly been very abundant in the Little Swanport estuary in the past, as shown by the massive pile of buried oyster shells at Seaford near the entrance to the estuary. This mound of oysters has been estimated to cover 32 ha and is up to 3 m deep (Dyke and Dyke unpublished 2002). According to the Register of the National Estate Database Place Report for Little Swanport Site, this is an Aboriginal midden of great scientific and historic significance. It is the most important archaeological site yet excavated in Tasmania and dates back to 2500 BC (DEWHA n.d.). These native oysters were also very abundant in most other estuaries of south-eastern Tasmania and were extensively harvested by dredging in the mid 1800s. The Royal Commission into the Fisheries of Tasmania in 1882 documented that 5.2 million oysters were marketed from the Swanport region in one of the best harvest years (Parliamentary Report 1883). These oysters were harvested by dredging the bottom and minimal consideration was given to managing the stock; juvenile oysters were generally dumped, burnt for lime, or used for road making. By the 1880s native oysters were in serious decline and although efforts were made to manage the fishery, it quickly collapsed and has never recovered. Dredge harvested native oysters across the southern and eastern coasts of mainland Australia have suffered a similar fate, and it is not generally recognised that the only known remaining oyster reef habitat of this species occurs in Georges Bay to the north of Little Swanport (Beck *et al.* 2011). Flat oysters do still occur in Little Swanport, but in limited quantities scattered across the sea bed. The loss of large beds of oysters and as a consequence their filter-feeding activity, which removed substantial quantities of microalgae and organic matter from the estuary, as well as the destructive harvesting practice of dredging the bottom, must have had a significant impact on the ecology of the estuary. Unfortunately there are no records on the ecology of the estuary in the early 1800s to compare with present day conditions.

Introduced species

Rice grass (*Spartina anglica*), a vigorous saltmarsh plant introduced into Tasmanian estuaries to assist with erosion control, spread across the mud flats in the upper estuary and was considered a problem as it encroached on the intertidal area. It affected the wildlife and reduced the area available for oyster farming, as well as altering the flow of water in the estuary. The estimated size of the infestation in 1997 was greater than 10 ha. A Tasmanian

Government initiative to eradicate or control rice grass invasions around the state has resulted in the eradication of rice grass beds from the estuary. Annual surveys are conducted to remove any seedlings arising from the seed bank deposited in the estuary; the number of seedlings removed is declining each year and in 2007 only a few plants were found.

The green crab, also known as the European shore crab (*Carcinus maenas*) is seasonally abundant in the estuary and has been found on oyster trays (Martin and Procter 2000). Several specimens of the introduced Japanese seastar (*Asterias amurensis*) have recently been found in the estuary (Lawler 2008). The Pacific oyster *Crassostrea gigas* is also an introduced species into the estuary. However, whether it is considered a benefit or a pest depends on the viewpoint of the individual.

Pathogen levels in the estuary

TASQAP monitors bacterial levels, toxic dinoflagellates and chemical residues in the Little Swanport estuary waters and shellfish as part of an internationally accepted program adopted by the Tasmanian government since the mid 1980s to reduce the food safety risks of shellfish consumption. The rationale of this program is that if shellfish are grown in clean, unpolluted waters they will be safe to eat.

This growing area was given an 'approved conditional' classification in 2001 because an adverse impact on water quality was observed after heavy rainfall, when the salinity dropped below 15 at sites near the leases. In the period 2001–10 the water quality of the estuary was consistently good at high salinities and low rainfall, but declined at relatively high rainfall levels. Several sites sampled did not meet the acceptable pathogen criteria (median <14 thermotolerant coliforms/100 mL and <10% samples with greater than 21 thermotolerant coliforms/100 mL) required for approved unconditional harvesting of shellfish. The 'approved conditional' classification is thus maintained (DHHS 2010). However, as we know, low salinities have become a rare event in Little Swanport and so this estuary is infrequently closed to shellfish harvesting. Heavy metal residues and other chemicals in shellfish were last tested in 2008 and showed no concentrations of concern. Little Swanport is also considered to be a low risk biotoxin area and so is sampled on a monthly basis. In 2007 low numbers of toxic dinoflagellates were recorded and no management action was required.

Future

The effects of climate change in Little Swanport are likely to be an increase in sea level and more varied rainfall. While global warming over the last 50 years has been attributed to human activities (Pittock 2003), the specific effects on climate rainfall patterns and sea level are complex and difficult to elucidate.

Rainfall records for the east coast of Tasmania over the last 100 years have shown a sudden drop in precipitation from 1980 to the present (Micevski *et al.* 2006). This reduction in rainfall has led to less freshwater input to the estuary, which in turn is likely to increase the influence of tidal marine input and the extent of hypersalinity in the upper estuary. It has also probably led to a slowing in the maturation of the estuary, as less sediment is transferred from the catchment with the reduced flow, although as yet there in no direct evidence for this. As this effect becomes more marked it is likely that components of the biota that are less able to handle the more consistently saline conditions will be affected.

11

Estuarine responses to environmental flows

Jeff Ross

We have already discussed the difficulties associated with the allocation of water across the Little Swanport catchment. Similar situations exist in many catchments across Australia, generally on a much larger and much more complex scale, such as in the Murray–Darling system. There, previous over-allocation of water in the headwaters and mid-stream sections has now resulted in catastrophic consequences downstream, especially in the Coorong and Lower Lakes near Adelaide. What is more, in areas such as Australia, climate change is predicted to alter the amount of freshwater available through changes to rainfall and associated run-off (Hughes 2003).

Considerable effort has been extended across Australia to better understand the dynamics of freshwater flow, largely to support the increased allocation of water to irrigation, industrial use and town water supplies. However, there have been few assessments of the importance of freshwater flows to estuaries, even though estuaries are receiving less and less water because of the increasing extraction upstream. The health of the estuarine environment is vitally important to most Australians because the majority of us live around and use estuaries for a variety of purposes – recreation, aesthetic value, transport, fishing, aquaculture, waste disposal. It is therefore essential to have some understanding of the consequences if freshwater flows are significantly reduced from their natural state.

Freshwater flows are an important supply of nutrients to estuaries and support the high productivity that occurs in these systems. We have already learnt that the Little Swanport estuary supports 42 species of fish and that six of the seven freshwater fish in the Little Swanport River move through estuaries to the sea at some stage during their life cycle. The oyster farmers in the lower estuary suggest that production of the Pacific oyster, *Crassostrea gigas*, is enhanced by freshwater flows into estuaries. This oyster is physiologically adapted to less than fully marine conditions, showing highest growth rates and survival at salinities of around 25. Additionally, inflowing water from the Little Swanport River delivers nutrients to the estuary that is likely to support increased production of phytoplankton, the small (often microscopic) photosynthetic algae drifting in the water column that are a major food source of oysters. However, the importance of these estuarine freshwater flows to shellfish production has never been assessed in Tasmania, and very little information is available from mainland Australia.

We have investigated the role of freshwater flows into the Little Swanport estuary for three main reasons.

1. We wanted to obtain a better understanding of how the estuary functions. How does it work, what are the important processes and key elements to making this a healthy estuary, what are the relative roles of water from the river and from the ocean, sources of essential nutrients, and key faunal and floral communities?
2. We wanted to assess the importance of freshwater flows to the overall function and productivity of the estuary. What will happen to the estuary if the total volume, timing of delivery and rate of delivery change?
3. We wished to examine the role of Pacific oysters in the estuary ecosystem. How much of the primary productivity in the estuary is consumed by the oysters, and how important are freshwater flows to the growth of oysters in the estuary?

To support our assessment of freshwater flows to the estuary we have used the data on the quality and flow of freshwater in the Little Swanport River collected by the Tasmanian Government and on estuarine water dynamics from research undertaken by TAFI, University of Tasmania, which have been described to some extent in chapters 9 and 10, respectively. These are expanded further here.

Three approaches have been undertaken to improve our knowledge of the role of freshwater flows to estuarine health and productivity. These are:

1. to look for empirical relationships based on field observations, such as between oyster growth rates and river flow
2. to develop a nutrient budget based on measurements to calculate annual inputs and outputs
3. to develop a dynamic ecosystem model capable of predicting the response of the estuary to freshwater flows on a day-by-day basis.

This component of the research has involved complex scientific assessment, which is summarised here rather than presented in detail.[1] Here we explain the information used for the analysis and the general principles behind the calculations and models.

Field observations

Arguably the most compelling method to determine the importance of freshwater flows for the health and productivity of the estuary is to use direct evidence; that is to observe if there is a direct relationship between freshwater flows and field observations of key estuarine indicators of health and productivity, such as the level of nutrients, biomass of phytoplankton, and oyster growth rates. In the most recent field study of Little Swanport estuary carried out by Crawford *et al.* (2006), monthly samples were collected at sites throughout the estuary between January 2004 and January 2005. Measurements included water column nutrients, chlorophyll a (as a proxy measurement for photosynthetic biomass in the water column), dissolved oxygen, turbidity, salinity, phytoplankton, zooplankton (the plankton that feeds on other plankton), and oyster growth. This work demonstrated that freshwater flows had a significant effect on salinity, turbidity, dissolved oxygen and nutrient levels in the estuary. However, monthly sampling did not provide the temporal resolution necessary to detect potential flow-on effects on the biology (e.g. effects on phytoplankton and zooplankton dynamics or oyster growth). To gain an improved understanding of the temporal dynamics of the estuary, including the response to freshwater flow, samples were collected weekly (chlorophyll a), fortnightly (nutrients and zooplankton) and bimonthly (oysters) between March 2006 and June 2008 in the lower estuary where the majority of oysters are farmed.

Nutrient budget

With the field observations in hand we then calculated the annual nutrient budget for the estuary: identifying and quantifying the relative importance of the major fluxes into (freshwater flows and ocean exchange) and out of the estuary (ocean exchange). The difference between the inputs and outputs indicates whether the estuary is a net sink (accumulator) or a net source (supplier) for the nutrient of interest. To ensure a common and consistent budgeting approach we used the Land Ocean Interactions in the Coastal Zone (LOICZ) program guidelines for the collection of empirical data and estimation of budgets (Gordon *et al.* 1996). These methods were applied to the Little Swanport estuary, approximated as a single well mixed box. The budget described here is for nitrogen, given that in most coastal marine ecosystems nitrogen is the key limiting nutrient (e.g. Boynton *et al.* 1982); that is, nitrogen is the nutrient that is in short supply relative to the others and will be exhausted first, thus limiting cellular growth. To help understand the influence of oyster aquaculture in Little Swanport, the role of oysters was also factored into the estuary's nitrogen budget. Table 11.1 lists the important reservoirs and internal and external fluxes that are elements of the nitrogen budget constructed for Little Swanport.

Reservoirs

To begin constructing the nitrogen budget, we required estimates of the nitrogen levels in phytoplankton (P), dissolved inorganic nitrogen (DIN), detritus (D) and oyster (O) reservoirs within the estuary. (Dissolved inorganic nitrogen is the sum of the concentrations of nitrate, nitrite and ammonia dissolved in the water, and detritus comprises the non-living particulate organic material, typically including dead organisms and fecal pallets.) To formulate the nitrogen budget, all quantities are expressed as an equivalent mass of nitrogen per unit volume, and later converted into t of nitrogen per year for the annual budgets.[2]

External fluxes

River loads

River inputs are based on river level data collected by DPIPWE at gauging station 2235, which is located approximately 1 km upstream of the upper limit of the estuary. The gauge's

Table 11.1. Elements of the nitrogen budget in Little Swanport estuary

Symbol	Description
Reservoirs	
P	Phytoplankton
DIN	Dissolved inorganic nitrogen
D	Detritus
O	Oysters
Oyster fluxes	
P → O	Ingestion
D → O	Ingestion
O → D	Biodeposition
O → Harvest	Oyster harvest
O → DIN	Excretion
External fluxes	
River	River loads
Ocean	Ocean exchange

river height–flow rating table was then used to generate daily river flow in ML/day. River load estimates were made using relationships between water quality parameters and river flow measurements at the gauge site. Physical and chemical data were also collected during monthly visits to the site and additional samples have also been collected during floods using automated sampling equipment. The relationship between dissolved inorganic nitrogen (DIN mg/L) and river flow (ML/day) was used to calculate river loads.[3] We calculated loads of detrital nitrogen from the river, but in this case there was no relationship between detrital nitrogen and river flow. A constant freshwater nitrogen concentration of 8 mg/L based on the average concentration measured by Crawford *et al.* (2006) was adopted. To provide an estimate of the instantaneous load for dissolved inorganic and particulate nitrogen, the concentration was multiplied by the discharge volume for that period.

Ocean exchange

Before we calculate the amount of DIN, P and D that moves across the estuary–ocean boundary, we need an estimate of their concentrations in both the ocean and the estuary (we already have an estimate of estuarine concentrations above), and the volume of water that exchanges across the estuary–ocean boundary in both directions.

Ocean concentrations

We used field data at the ocean reference site from the most recent study (Crawford *et al.* 2006) for setting ocean boundary conditions for DIN, P and D. For P (measured as chlorophyll a) there were very few data points and no apparent seasonal patterns, so we have adopted a constant ocean boundary condition of 7 mg N/m^3, which is the average concentration measured by Crawford *et al.* (2006). For detrital N there were negligible amounts recorded, and thus we adopted a constant boundary condition of 0 mg N/m^3.

Exchange volumes

In systems (in our case the Little Swanport estuary represented as a single box) for which freshwater inflow and salinity data are available, and if there is a difference in salinity between the system of interest and adjacent waters, a simple water and salt budget can be used to determine water exchange between the system and adjacent waters. In this study, the fraction of freshwater method outlined by Dyer (1973) was used to calculate the amount of time that freshwater spends in the estuary, commonly referred to as the 'flushing time' (t_f). With the Little Swanport estuary approximated as a single well mixed box, flushing time of the estuary is defined as:

$$t_f = \frac{\text{freshwater volume}}{\text{freshwater input}} = \frac{\left(\dfrac{S_{oc} - S_{est}}{S_{oc}}\right) \cdot V}{Q_f}$$

where S_{oc} = ocean salinity

S_{est} = estuarine salinity

V = estuarine volume

Q_f = freshwater input.

Given the importance of river flow as a determinant of flushing time and hence water exchange across the mouth of the estuary, the relationship between river flow and flushing time was fundamental to calculating budgets in an estuary with highly variable river flow.

Using salinity measurements collected under different river flow conditions we developed a model that resolves variations in exchange volumes (and the substances it contains) across the estuary ocean boundary that arise from daily variation in river discharge (Figure 11.1). The relationship between flushing time and river flow for the Little Swanport estuary is a power function: as river flow increases, flushing time decreases quickly, but at high river flows flushing time is relatively stable. In the context of estuarine hydrodynamics, the most probable explanation for this relationship is best described in three phases. When river flow is absent or negligible, tidal energy is the major driver of physical circulation and exchange with the ocean. As river flow increases, the flow of freshwater causes a characteristic circulation pattern (commonly referred to as 'estuarine circulation') in which the freshwater entering at the head of the estuary creates a near-surface flow of lighter freshwater out of the estuary and a compensating deep flow of saltier water up towards the head of the estuary.[4] This buoyancy-driven circulation enhances exchange with the ocean, and the magnitude of this exchange increases with river flow. However, when river flow is sufficiently large, it so dominates exchange with the ocean at the mouth of the estuary that there is little influence of tidal energy and no compensatory deeper flow of saltier water from the ocean to the estuary. For the Little Swanport estuary, it appears that estuarine circulation breaks down and river flow dominates physical exchange with the ocean when river flow is around 500–1000 ML/day.

The relationship between flushing time and river flow is then used to calculate the exchange of material across the estuary–ocean boundary.

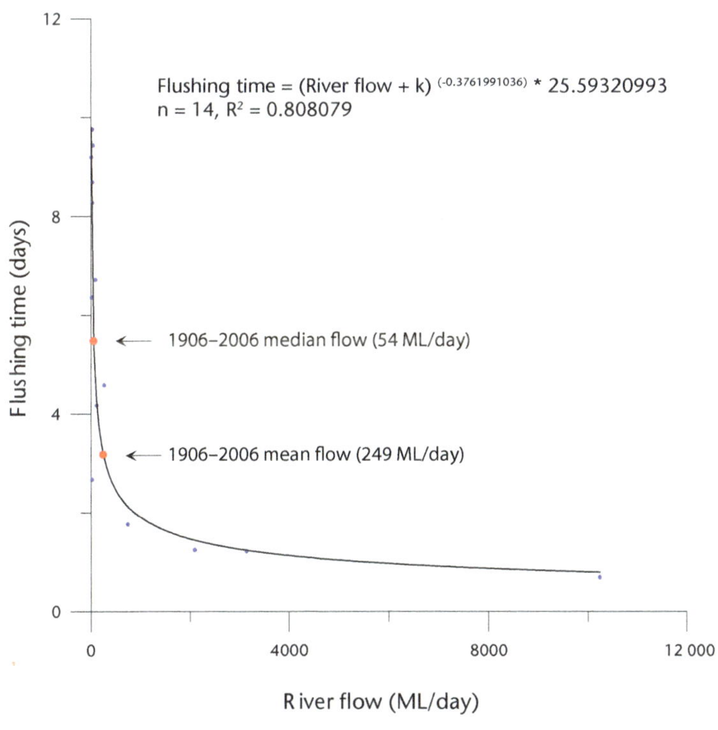

Figure 11.1 The relationship between flushing time and river flow for the Little Swanport estuary. The small dots show the calculated flushing times for field measurements and the solid line represents the negative power function fitted to this data. The larger dots represent the mean and median flows of the daily flow time series over 100 years of current water use generated from a water balance model established for the catchment by Sinclair Knight Merz (2004).

Oysters

To help factor the role of oysters into the annual nutrient budgets, the amount of phytoplankton (P) and detrital nitrogen (D) consumed by oysters was calculated. This depends on the rate at which oysters filter water (i.e. their clearance rate) and the concentrations of P and D in the water column. Although clearance rates (L/h) are likely to increase with the size of oyster, Crawford *et al.* (1996) found no relationship with size when expressed as clearance rate per dry tissue weight. The clearance rate for oysters used was 43.16 L/g dry weight/day. This is within the range of values observed for *Crassostrea gigas* in oyster farms in Tasmania by Crawford *et al.* (1996). The amount of P and D ingested was calculated by multiplying the clearance rate by their respective concentrations in the estuary. This was multiplied by the estimated biomass of oysters (in dry weight) in the estuary to calculate the total amount of nitrogen removed.

Of course, not all food consumed by the oysters is assimilated into new oyster tissue and a proportion is excreted as DIN and produced as faeces (i.e. detritus). Assimilation efficiency, the proportion of food ingested that is actually used for growth, was given a value of 0.5 for phytoplankton, based on the experiments of Crawford *et al.* (1996). Because filter feeders such as oysters often assimilate detritus much less efficiently than they do phytoplankton (e.g. Wilson *et al.* 1993), the assimilation efficiency for detritus was given a value of 0.25. We then assume that 25 per cent of the phytoplankton and detrital nitrogen assimilated is excreted as DIN. The amount of nitrogen produced as faeces (i.e. detritus) by the oysters is the difference between the nitrogen ingested and the nitrogen assimilated.

Ecosystem box model

As we discussed in the previous section, construction of an observation-based nitrogen budget is the simplest approach to quantifying the important fluxes into and out of the system; however, in systems with highly episodic river flow such as Little Swanport, field sampling regimes are normally not conducted at high enough frequency to understand the discharge variation and the estuarine responses that they may cause (see Webster *et al.* 2000). To resolve variations in estuarine responses that can arise from daily variation in river discharge, an ecosystem box model was developed for the Little Swanport estuary. Perhaps most importantly, the catchment was in drought during this study, and as such, we were left with no other option than to use a model for predicting estuarine response to different river flow scenarios. The model was developed using the STELLA software package (www.iseesystems.com).

Model structure and behaviour

An ecosystem box model is a mathematical description of an ecosystem in which the ecosystem is represented as one or more boxes. The basic structure of the model describes the ecosystems of the coastal embayments of south-east Australia (see Figure 11.2 for a schematic diagram of the model structure). Because the Little Swanport estuary appears to be well mixed it was treated as a single well mixed box. The transport of water column properties across the estuary–ocean boundary was calculated based on the model of river flow versus flushing time described above.

This model was written as a nitrogen model because, as explained, there is widespread evidence that nitrogen is the key limiting nutrient in coastal marine ecosystems. Therefore, all reservoirs (boxes in Figure 11.2) and fluxes (arrows in Figure 11.2) in the model are represented as mass of nitrogen per unit volume and mass of nitrogen per unit volume per unit time. The model structure and behaviour was verified by comparison with documented empirical responses to altered forcing functions, such as nutrient loads (e.g. phytoplankton biomass response). The model was then calibrated and validated for Little Swanport by comparisons with field observations.[6] It must be emphasised that the range of conditions during the calibration and subsequent validation periods were similar, particularly in regard to river flow, and thus the good fit during the validation period confirms the model behaviour under periods of low river flow. Ideally, it would be preferable to validate the model against observations obtained from a period in which other river conditions prevail from those in the period of data collection used for the calibration. Nonetheless, in the absence of this data, the model can still be used as a management tool for predicting estuarine response to different river flow scenarios. As more data becomes available under different river flow conditions, this model will be refined further, ensuring greater certainty in predictions.

Model forcing

The model is driven by inputs of nutrients from the river, by exchanges of nutrients, phytoplankton and zooplankton at the boundary with the ocean, and by seasonal variation in temperature and solar irradiance. River flow and associated nutrient loads were estimated on a daily time step as described above for the nutrient budget. Similarly, ocean boundary conditions and exchanges across the ocean boundary were calculated on a daily time step as described above for the budget.

The model includes three sub-models: a water column model, an epibenthic model and a sediment model.

The water column model is based on a nitrogen-phytoplankton-zooplankton-detritus model or so-called NPZD model (e.g. Fasham 1990), which has been extended to include oysters (O) that are suspended in the water column on oyster racks. Detritus has also been divided into detritus that is easily broken down or labile (DL) and refractory (DR) particulate detritus that is broken down with difficulty, and refractory dissolved organic matter (DON). Nitrogen, as DIN, DON, DL and DR, enters the water column model via river discharge, and, together with P and Z, DIN, DON, DL and DR are also exchanged with the ocean across the estuary mouth. Internally, DIN is taken up during phytoplankton growth; phytoplankton is consumed by zooplankton; detritus is generated through zooplankton and phytoplankton mortality and grazing; and DIN is generated through zooplankton respiration or mineralisation of detritus (either directly or via DON). Oysters consume

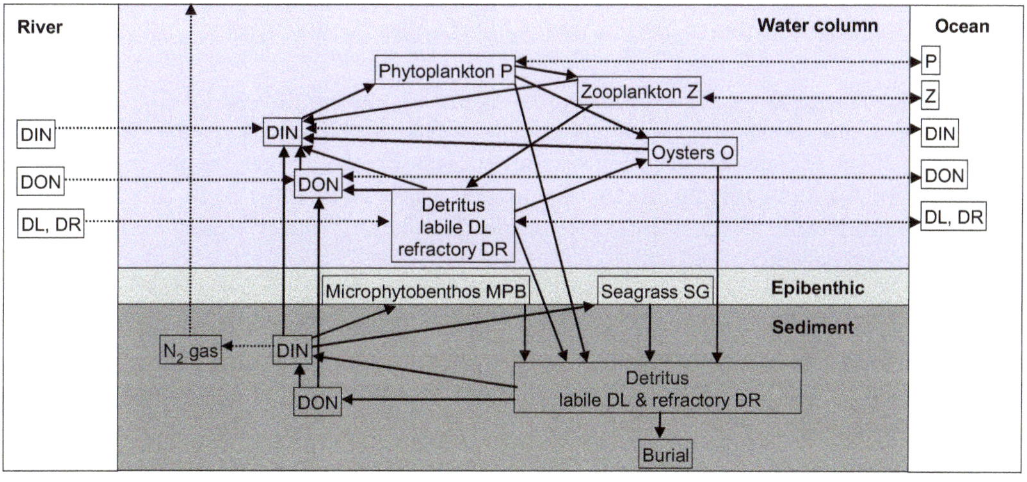

Figure 11.2: Schematic diagram of the model structure.

phytoplankton and detritus, and generate DIN and detritus through excretion and fecal production, respectively. Oyster biomass is held constant throughout the year based on industry estimates, with oyster growth removed during harvest. The biomass value takes into account industry estimates of mortality.

The epibenthic model represents benthic plants (seagrass (SG) and microphytobenthos (MPB; small photosynthetic algae that live on the seabed)) located at the sediment–water-column boundary. The benthic plants grow subject to light and nutrient conditions at the sediment surface. The sediment model includes the bacterial breakdown of particulate and dissolved organic matter derived from the water column and sediments to dissolved inorganic nitrogen (DIN). The subsequent rates of nitrification (conversion of ammonia to the nitrates and nitrites) and denitrification (loss of nitrogen from the system as nitrogen gas) are represented implicitly, with their relative importance dependent on the overall sediment respiration rate.

Model applications

In the first instance, the model was used to calculate annual pools (the amount of nitrogen found in each reservoir over a year) and fluxes (the amount of nitrogen that moves between reservoirs each year) for comparison with the results of the observation-based nutrient budget, and to provide estimates of the other reservoirs and fluxes that could not be directly measured or estimated. The model was then used to examine the role the river plays in the functioning of the estuary, and the likely impacts of changes in river flow to the estuary. The model was applied to the Little Swanport estuary under three scenarios.

1. Although the estuarine response to river flow is no doubt likely to vary depending on the magnitude, frequency and history of river flow events, in the first instance the model was used to examine the role of base river flow to the estuary. Model simulations were carried out for a two-year period, and the outputs compared across simulations with different base flows, ranging from 0 to 200 ML/day/year. This helps us asses the relevance of the cease-to-take flow periods (≤7.6 ML/day, November to April, and <9.5 ML/day, May to October) in the plan that are ostensibly based on the environmental water requirements of the Little Swanport River rather than the estuary.
2. Over the course of this study (and the study by Crawford *et al.* (2006)), the catchment moved into a drought. In 2004 and 2005, total flow into the estuary was ~31 251 and 75 258 ML respectively, compared with 1238 and 4258 ML in 2006 and 2007 respectively. To understand the impact of the drought on the estuary, and to help gain a better understanding of the influence of natural variability in river flow, the model was run for the 2004–2007 period.
3. As part of the Little Swanport Catchment Water Management Plan (DPIW 2006a), the catchment allocation limit was increased from 3882 to 6084 ML per year. To assess the impact of the increased allocation on the estuary, the model simulations using hydrographs of the flow regime under the previous allocation limit were compared with model simulations using hydrographs under the plan. This was repeated in a dry year and an average year.

Results and discussion

Field observations

With the exception of a few very small flow events in early to mid 2007 there was virtually no river flow into the estuary during the study period (Figure 11.3a). While this clearly limited our ability to collect empirical evidence on the effects of river flow on estuarine dynamics, it provided an opportunity to understand how the estuary functions when river inputs are negligible. Dissolved inorganic nitrogen (DIN) concentrations varied considerably over the two-year period, with no clear seasonal patterns (Figure 11.3b). However, in terms of the major constituents of DIN – ammonia and nitrate – ammonia largely tracked and was responsible for the pattern in DIN concentrations, but nitrate displayed a distinct seasonal pattern with concentrations increasing to 10–20 mg N/m^3 in the middle of each winter from a very low background concentration of less than 4 mg N/m^3. This reflects a similar pattern seen in the ocean outside the estuary, and, as such, the ocean may be the source of the winter nitrates observed in the estuary; however, see below.

The elevated nitrate concentrations on the east coast of Tasmania in winter are due to the influence of nutrient-rich sub-Antarctic waters. In contrast, concentrations are lower between autumn and spring due to the influence of the nutrient-poor East Australian Current (EAC) and increased biological activity and uptake of nutrients (Clementson *et al.* 1989; Harris *et al.* 1987). Phytoplankton, measured as chlorophyll a, also showed a distinct seasonal pattern, but with concentration peaking in late summer before declining to a minimum in late winter–spring (Figure 11.3c). A very similar seasonal pattern was recorded by Brown and McCausland (1999) during fortnightly sampling between 1992 and 1997 further up the estuary at the oyster nursery. Interestingly, in the current study the maximum phytoplankton biomass occurred just before rather than after the nitrate peaks, indicating that the nitrate peaks may be due to the senescence and subsequent remineralisation of the dead plankton rather than to nitrate imported from the ocean. Zooplankton biomass also peaked in summer (Figure 11.3d). The increase in zooplankton biomass appeared to be slightly later and out of phase with phytoplankton, as we might expect if the zooplankton are consuming the phytoplankton. However, the zooplankton peak was much shorter lived than that of the phytoplankton, raising the possibility that predators of zooplankton, such as fish, may be controlling their biomass.

Oyster growth, measured as milligrams of nitrogen gained per milligram N dry tissue weight (DTW) per day, hovered between a 0.2–0.4 per cent increase per day in DTW over the two-year period, with no apparent seasonal pattern (Figure 11.3e). On the other hand, the increase in oyster condition in each two-month period was reduced in the middle of summer each year, and the oysters actually lost condition in summer 2007–08 (Figure 11.3f). The loss of condition in summer is likely to be a direct result of summer spawning. This appears to be broadly consistent with the annual reproductive cycle of *C. gigas* that has been widely described (e.g. Dinamani 1987; Ren *et al.* 2003). In temperate regions, *C. gigas* exhibits a seasonal reproductive cycle that is clearly related to temperature, with initiation of gametogenesis in winter when water temperature is low, an active phase of gametogenesis (growing stage) in spring when water temperature has increased, maturity and spawning in summer when temperature is above 19°C (Mann 1979), and a resorption period in autumn (degenerating stage).

Nitrogen budgets

Observation-based budget

Several key aspects of the nitrogen cycle in Little Swanport are apparent from the estuarine-wide annual average pools and fluxes of nitrogen, particularly when comparing the relatively wet 2004 with the very dry years of 2006 and 2007 (Figure 11.4). The estimated river load of dissolved inorganic nitrogen to the estuary varied between approximately 113 t in 2004, 30 kg in 2006 and 200 kg in 2007. The corresponding load of detrital nitrogen was approximately 2.1 t in 2004 compared with 10 kg and 40 kg in 2006 and 2007 respectively. However, it is important to note that the accuracy of the annual load and flux

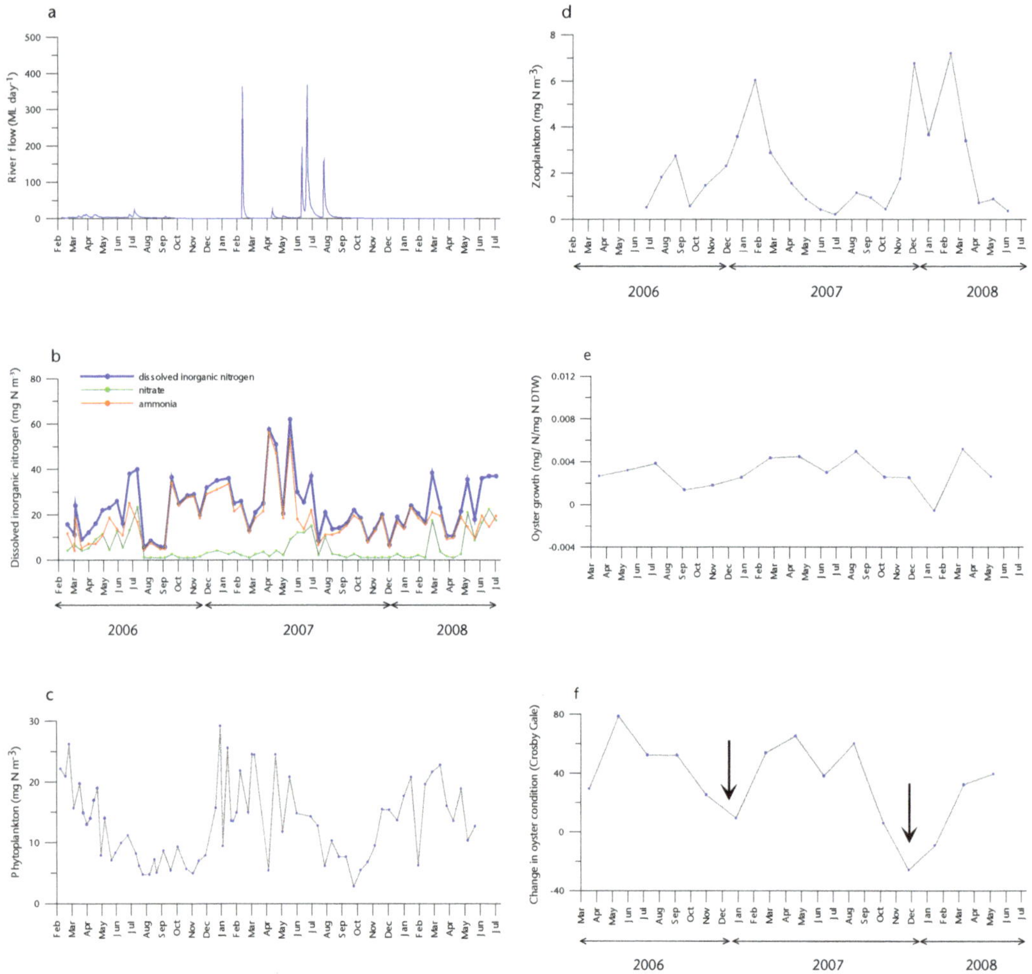

Figure 11.3 Time series plots of (a) river flow, (b) dissolved inorganic nitrogen, ammonia, nitrate, (c) biomass of phytoplankton and (d) zooplankton, (e) oyster growth, and (f) oyster condition (Crosby and Gale 1990). The oyster measures are based on changes during consecutive two-month deployments. Arrows represent likely time of spawning.

estimates is directly related to the representiveness of the sample days used to calculate the budget. Webster *et al.* (2000) demonstrated the large errors that can occur if the sampling regime does not represent the time variant nature of the discharge loads. For this reason we suggest that the budget for 2004 is likely to be overestimated. In 2004, the January sample coincided exactly with the peak of a one-in-five-year flood. With one sample day per month, the calculation essentially assumes that the conditions on that day prevail for the whole month, which is clearly not the case for floods of this magnitude. This is perhaps best described in terms of river flow; the 12 sample days used in the 2004 budget provide an estimate of 273 826 ML for the year compared with 28 295 ML when the January sample is excluded. The actual total river flow measured for the year was 31 250 ML. Nonetheless, there is still at least a five-fold greater load in 2004 compared with 2006 and 2007, with or without the January sample included (Figure 11.4).

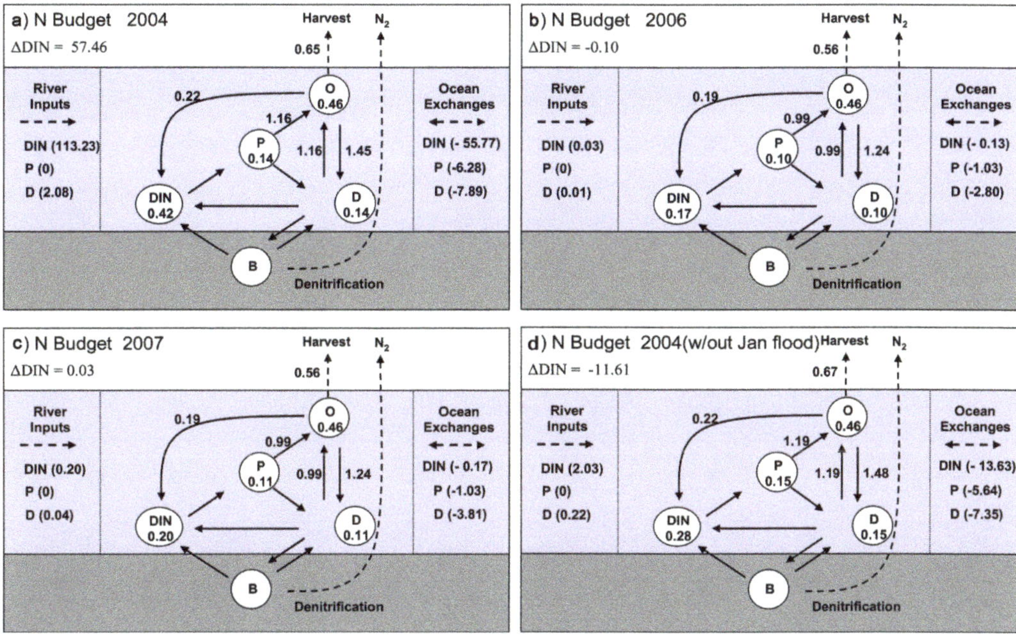

Figure 11.4 Annual nitrogen budgets based on field observations. The budgets are calculated for (a) 2004, (b) 2006 and (c) 2007 when the total river flow into the estuary was 31 251, 1238 and 4258 ML respectively; (d) represents the 2004 budget when the sample corresponding to a major flood in 2004 was excluded. Solid arrows represent internal fluxes, and dashed arrows are external inputs and outputs. Reservoirs are annual averages (t N) and all other quantities are annual fluxes (t N year^{-1}).

The importance of oceanic exchange, and in particular the influence of river flow on the magnitude of this oceanic exchange, is also clearly evident. In 2004, there is a large export to the ocean because a significant fraction of the river input is washed directly out to sea during large river flows. In 2006 and 2007, oceanic exchange also led to a net export, albeit a much smaller one. The net export in all three years is largely a reflection of the higher annual average concentrations in the estuary compared with the ocean. Despite the increase in exports to the ocean when river flow was greatest in 2004, not all DIN constituents were flushed to the sea, and the standing stock of DIN, phytoplankton and detritus in the estuary were greater than 2006–07 as a result. The concomitant increase in oyster harvest was estimated at approximately 100 kg N, which equates to an increase of approximately 15 per cent in 2004 relative to 2006 and 2007.

Ecosystem model-based budget

To gain insight into the other major pools and fluxes, which could not be directly estimated in an observation-based nitrogen budget, the annual pools and fluxes estimated by the ecosystem model were compared. In general, the model reproduces pools and fluxes similar to those of the observation-based budget. The major difference was the higher estimates for the oyster harvest and biomass of detritus in the water column in the observation-based budget. However, it is important to note that there were few instances when phytoplankton biomass and detrital biomass were measured concurrently, and thus, the 50:50 split assumed in the observational budget (see Crawford *et al.* 2010 for details) should

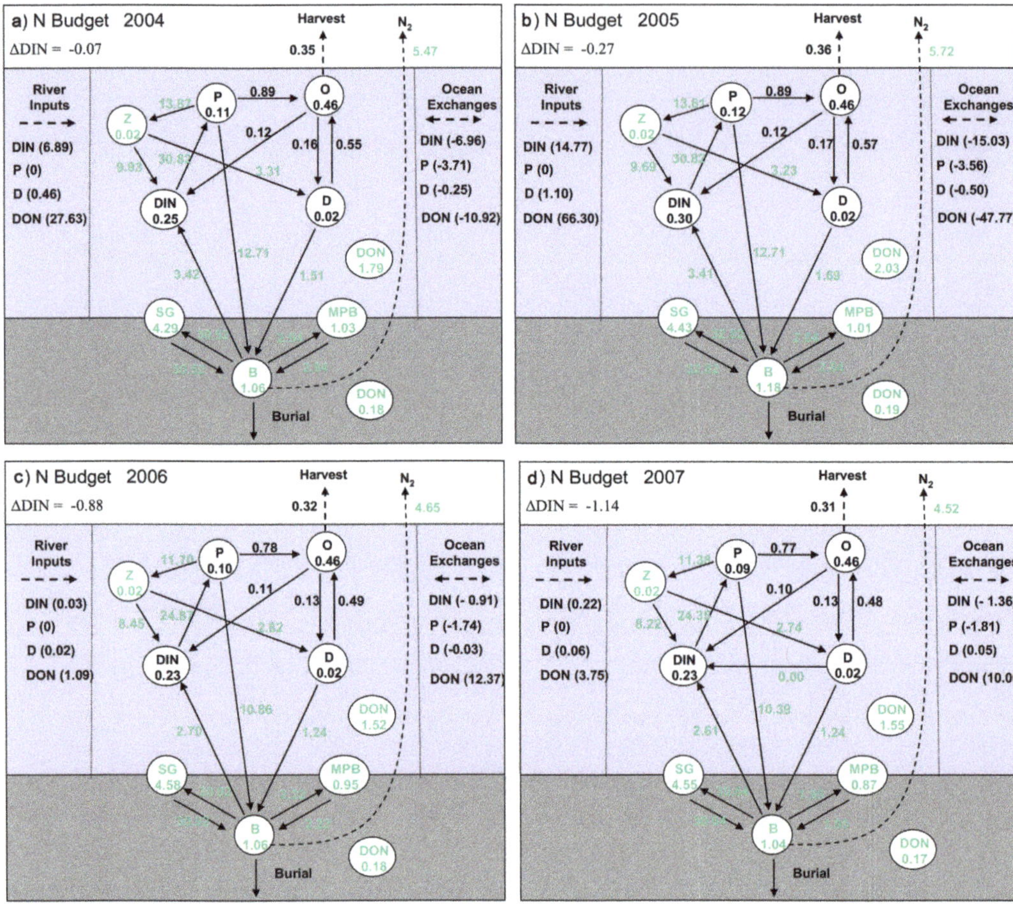

Figure 11.5 Annual nitrogen budgets based on the ecosystem model. The budgets are calculated for (a) 2004, (b) 2005, (c) 2006 and (d) 2007 when the total river flow into the estuary was 31 251, 75 258, 1238 and 4258 ML respectively. Solid arrows represent internal fluxes and dashed arrows are external inputs and outputs. Reserviors are annual averages (t N) and all other quantities are annual fluxes (t N year^{-1}).

be treated with caution. Irrespective of the budget method used, oyster harvest, DIN, P and D pools were higher in the wet years of 2004 and 2005 compared with dry years of 2006 and 2007. With the exception of seagrass production, rates of primary (phytoplankton and microphytobenthos) and secondary production (zooplankton and oysters) were also higher in wet than in dry years.

There were a number of other key aspects of the nitrogen cycle in Little Swanport apparent from the model. The flux of nitrogen through the primary producers – seagrass, phytoplankton and microphytobenthos – far outweighs the external inputs of N to the estuary due to the efficient internal recycling in the sediments and water column. Although the phytoplankton pool is very small compared with that of seagrass and microphytobenthos, phytoplankton production is very high due to its rapid turnover rate. In contrast, the seagrass and microphytobenthos pools are much larger, but turn over much more slowly. Secondary production in the form of zooplankton growth also represents a major flux of nitrogen despite the small zooplankton pool. In contrast, the flux of nitrogen through oysters is comparably small, despite representing a larger pool of nitrogen in the water column than phytoplankton and zooplankton.

Of the phytoplankton primary production, the model estimates that approximately 46 per cent sinks to the sediment, 42 per cent is consumed by zooplankton, 3 per cent is consumed by oysters, and 10 per cent is exported to the ocean. Although phytoplankton primary productivity was higher in the wet years than the dry, a greater proportion of the primary productivity generated in the dry years was recycled internally. For example, in the dry years of 2006 and 2007 there was a 2–3% increase in the proportion of the phytoplankton productivity recycled internally via zooplankton grazing and sedimentation. This is because the average flushing time is longer (and hence the chance of being exported to the ocean is reduced) in the dry (~11 days) compared with the wet years (~8 days). In the wet years approximately 12 per cent of phytoplankton productivity is exported to the ocean, compared with approximately 7 per cent in the dry years. The dominant form of nitrogen in the river inputs and water column in the estuary is dissolved organic nitrogen (DON). Not surprisingly, DON also accounts for most of the nitrogen fluxing across the ocean boundary. In the wet years of 2004 and 2005 when estuarine concentrations were high and flushing times were reduced, a large proportion of the river load was exported to the ocean, but in the dry years of 2006 and 2007 when estuarine concentrations were reduced, there was a net import of DON from the ocean. Of the organic nitrogen deposited from the water column that is fluxed through the sediments, denitrification removes approximately 40 per cent.

Ecosystem box model

Effects of base river flow on estuarine dynamics

A major goal of the study was to understand the importance of river flow to the functioning of the estuarine ecosystem. To gain a general understanding of the role of base river flow to the estuary, model simulations were carried out for a two-year period, and the outputs compared across simulations with different base flows, ranging from 0 to 200 ML/day/year (Figure 11.6). Over this range, there is a nonlinear increase in the average phytoplankton biomass with increasing base flow. Phytoplankton biomass increases rapidly with base flow from 0–20 ML/day, before the rate of increase slows to a more steady rate as base flow increases over 40 ML/day (Figure 11.6a). Not surprisingly, given that oysters feed on phytoplankton, oyster growth shows a similar response to increasing base flow (Figure 11.6b). These nonlinear responses are driven largely from the nonlinear relationship between flushing time and river flow that was determined when developing the transport model (Figure 11.1). At low river flows, phytoplankton has more time to take up the additional nutrient inputs from the river because the flushing times with the ocean are long. As river flow increases, this uptake obviously drops off as flushing time shortens. In other words, the greatest benefits per ML of river flow for phytoplankton production and oyster growth are at low base flows; increasing base flow from 1 to 40 ML/day leads to an approximately 11 per cent increase in oyster growth rates and it isn't until the base flow reaches 200 ML that another increase in growth of approximately 11 per cent is reached.

In contrast to phytoplankton biomass and oyster growth, zooplankton decreases with increasing base flow (Figure 11.6c). This is largely because zooplankton have a slower turnover rate and any increase in biomass is countered by increased export to the ocean as a result of the increase in flushing time with river flow. Although oysters have even slower turnover rates, they are not subject to being flushed from the estuary. Similarly, microphytobenthos, which live for the most part on the sediment surface, are not likely to be

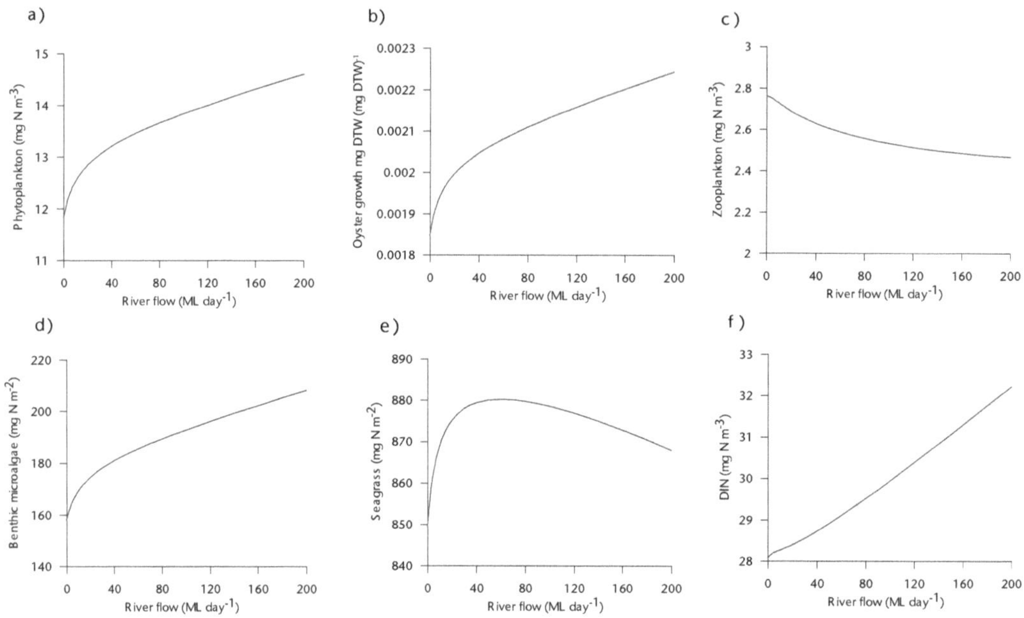

Figure 11.6 Effects of base river flow to the estuary. Base flow was kept constant for a two-year period, and the average outputs compared across simulations with different base flows, ranging from 0 to 200 ML/day/year.

susceptible to being flushed from the estuary at this range of base flows, explaining why their biomass also increases with base flow, particularly at low flows (Figure 11.6d).

Interestingly, seagrass, the other major benthic primary producer, also increases in abundance at low flows, but declines in abundance once the base flow reaches approximately 60 ML/day (Figure 11.6e). This is because seagrass is susceptible to overgrowth by epiphytes when nutrient concentrations are high (e.g. Madden and Kemp 1996). Although epiphytic organisms growing on or attached to the seagrass are not modelled explicitly, they are represented as an additional specific loss rate for seagrass that is proportional to DIN concentrations in the water column, which increase in concentration with base flow (Figure 11.6f).

Given the seasonal patterns observed for many of the major nitrogen reservoirs in the model simulations, the movement of nitrogen into and out of the estuary that underpins the net response discussed above is also going to vary over time. This is because the amount of N that moves across the estuary–ocean boundary depends on the river flow (and hence flushing time) and the size of the difference between the ocean and estuarine concentrations, and the direction of the movement depends on whether the ocean or the estuary has the highest concentration. For example, from January to June, DIN concentrations are lower in the estuary than the ocean and there is a net import of DIN into the estuary from the ocean, but for the remainder of the year the concentration is higher in the estuary and there is a net export of DIN from the estuary. The magnitude of these fluxes is greater when base river flow is 100 ML/day compared with 10 ML/day because the flushing time is shorter and hence the volume of water exchanged with the ocean is greater. In the case of phytoplankton and with the exception of spring, the estuary is a net exporter to the ocean, particularly in late summer when concentrations in the estuary are at their peak.

These results also suggest the simple hypothesis that the greatest benefits of river flow to phytoplankton production will come from flows in winter months when the difference in ocean and estuarine concentrations, and hence loss to the ocean, is at its minimum. As it stands, the Water Management Plan for the Little Swanport catchment has 'cease-to-take' periods that allow for greater environmental flows in winter than in summer (£7.6 ML/day, November to April, and £9.5 ML/day, May to October). Ostensibly based on the environmental water requirements of the Little Swanport River rather than the estuary, our simple hypothesis for the estuary suggests that this will also have maximum benefit for estuarine productivity. To test this the model was re-run, comparing the estuarine response when all river flow was in winter (100 ML/day from June to October) versus when all river flow was in summer (100 ML/day from November to March). Interestingly, the reverse was actually true, with greatest productivity achieved following summer rather than winter river flows. This is because although losses to the ocean were reduced with winter rather than summer flows, growth rates of phytoplankton are limited by temperature in winter and there was virtually no detectable increase in biomass. In contrast, during the summer flows increases in production in the estuary far outweighed the increase in losses to the ocean.

Effects of the drought on estuarine dynamics

One of the most powerful tests of the importance of freshwater flows to an estuary would be to compare years at each end of the potential extremes; times characterised by high flows compared with periods characterised by very low or no flows. If differences in the health and productivity of the estuary are not evident between these extremes then it would seem unlikely that more subtle changes are likely to be significant. Therefore, a comparison of model outputs for the two drought years of 2006 and 2007 (when the total flow into the estuary was 1238 and 4258 ML respectively) with the two years prior to this, 2004 and 2005 (when the total flow into the estuary was ~31 251 and 75 258 ML respectively), provided a critical test of the importance of freshwater flows to the Little Swanport estuary. Model outputs for some of the key state variables and important fluxes are shown in Figure 11.7. It is clear that the major river flow events in 2004 and 2005 led to an increase in DIN concentrations in the estuary and subsequent phytoplankton blooms in the estuary. The increase in phytoplankton biomass also led to an increase in zooplankton biomass, albeit of a smaller magnitude. As we discovered in the base flow comparison above, the higher estuarine DIN concentrations following the floods led to a decline in seagrass biomass due to epiphyte growth on seagrass fronds. There was also a response by benthic microalgae to the floods (Figure 11.7); however, this was far more subtle than that of the other primary producers, particularly phytoplankton. This is likely to reflect a combination of the slow turnover rates and dependence on DIN in the sediments by benthic microalgae.

As we might expect, the increased production of organic matter following the floods led to a concomitant increase in the supply of organic matter to the sediments as evidenced by the response in rates of sediment denitrification and release of DIN back into the water column.

In gross terms, when the responses shown in Figure 11.7 are integrated over the respective two-year periods, the model predicts (Table 11.2) that the 95 per cent reduction in river flows from 2004–05 to 2006–07 led to a 15 per cent reduction in estuarine DIN concentrations and phytoplankton biomass, an 11 per cent reduction in benthic microalgae biomass, a 12 per cent reduction in oyster harvest, a 1 per cent reduction in zooplankton biomass, and a 5 per cent increase in seagrass biomass.

Effects of increased allocation on estuarine dynamics

As part of the Water Management Plan (DPIW 2006a) the catchment allocation limit (the total amount of the catchment's water resource that can be allocated for stock and domestic

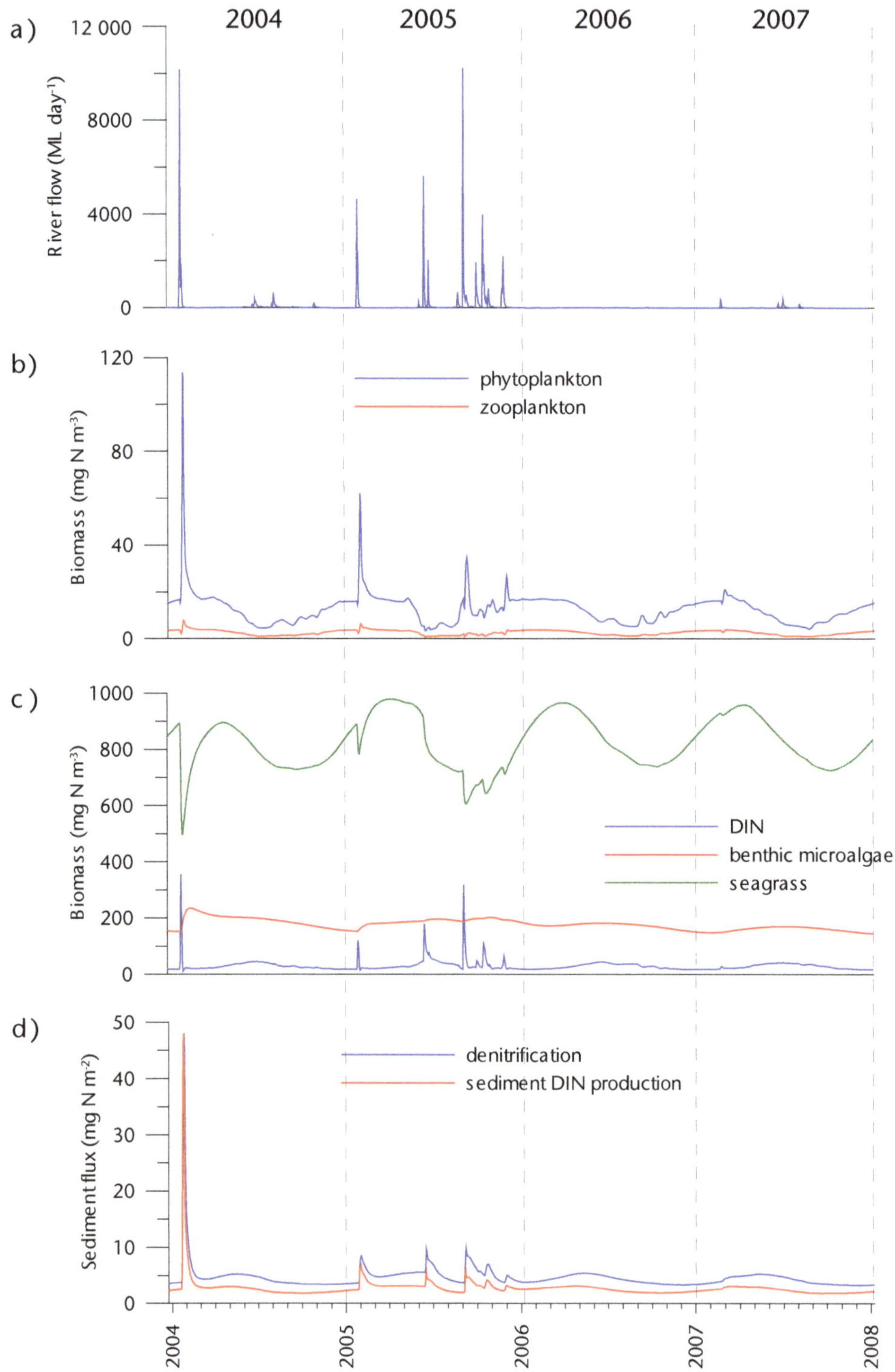

Figure 11.7 Time series plot of (a) river flow into the estuary, and the model responses for (b) phytoplankton and zooplankton biomass, (c) dissolved inorganic nitrogen concentration in the water column, benthic microalgae and seagrass biomass, and (d) rates of sediment denitrification and dissolved inorganic nitrogen release from the sediments.

Table 11.2. Comparison of total oyster harvest, average biomass of phytoplankton, seagrass, MPB and zooplankton, and the average concentration of DIN in the estuary from 2004 to 2007.

Year	River flow (ML)	Oyster harvest (kg)	Phyto-plankton (kg)	DIN (kg)	Zoo-plankton (kg)	MPB (kg)	Seagrass (kg)
2004	31 361	350.9	109.6	251.9	22.0	1034.2	4288.1
2005	75 258	362.2	116.1	295.0	22.3	1009.5	4433.3
2006	1238	316.1	97.4	232.9	22.1	948.0	4582.1
2007	4258	310.5	94.9	231.9	21.5	870.2	4549.1
Change (04–05 vs. 06–07)	50 562	43.3	16.7	41.1	0.3	112.7	–204.9
Change as percentage of 04–05	–95	–12	–15	–15	–1	–11	5

or irrigation purposes) was increased from 3882 to 6084 ML/year. To assess the impact of the increased allocation on the estuary, the model simulations using hydrographs of the flow regime under the previous allocation limit were compared with model simulations using hydrographs under the plan. This was repeated for a dry year and in an average year. The model outputs for both allocation scenarios were barely distinguishable, as evidenced in Tables 11.3 (a dry year) and 11.4 (an average year). These compare the total oyster harvest, the average biomass of phytoplankton, seagrass and zooplankton, and the average concentration of DIN in the estuary under the different allocations.

River flow in the drought years of 2006 (1238 ML) and 2007 (4258 ML) was significantly lower than in the dry year scenario (~20 000 ML) tested above, and as such we assessed the effects of the increased allocation in a 'very dry' year (2007). In contrast to

Table 11.3. Total oyster harvest, average biomass of phytoplankton, seagrass, and zooplankton, and the average concentration of DIN in the estuary under current and full uptake (under the plan) conditions in a dry year.

	Current	Full	Change	%
Oyster harvest (kg)	343.480	343.159	0.321	0.000 002
Phytoplankton (kg)	106.718	106.611	0.107	0.000 015
Seagrass	7156.356	7158.753	–2.397	0.000 017
DIN (kg)	236.128	235.513	0.615	–0.000 027
Zooplankton (kg)	21.083	21.171	–0.087	–0.000 033

Table 11.4. Total oyster harvest, average biomass of phytoplankton, seagrass, and zooplankton, and the average concentration of DIN in the estuary under current and full uptake (under the plan) conditions in an average year.

	Current	Full	Change	%
Oyster harvest (kg)	361.585	361.503	0.081	0.000 002
Phytoplankton (kg)	115.589	115.600	–0.011	0.000 015
Seagrass	6630.757	6668.850	–38.092	0.000 017
DIN (kg)	288.656	286.294	2.363	–0.000 027
Zooplankton (kg)	20.828	20.896	–0.068	–0.000 033

Table 11.5. Total oyster harvest, average biomass of phytoplankton, seagrass, and zooplankton, and the average concentration of DIN in the estuary under current and full uptake (under the plan) conditions in an average year.

	Oyster (kg N)	Phytoplankton (kg N)	DIN (kg N)	Zooplankton (kg N)
2007 – current	325.1	101.1	223.5	22.5
2007 – full uptake	320.8	100.3	223.0	22.6
Change current to full uptake (%)	1.31	0.73	0.20	–0.18

previous cases, there was a discernable difference between the current and full allocation scenarios in the very dry year scenario. The model predicted that the reduced load of nutrients to the estuary in the February flow event, due to likely water harvesting under the full uptake conditions, would lead to a decline in estuarine nutrient concentrations and a subsequent repression of the phytoplankton and oyster growth responses. However, in winter, when estuarine nutrient concentrations are already higher and growth rates are limited by temperature, the model predicts that harvesting flows of a similar magnitude will have a negligible effect on phytoplankton and oyster growth.

Taken across the whole year, the changes due to full allocation are relatively small (Table 11.5). Changes of this magnitude fall within the uncertainty inherent in model simulations and, as such, should be treated with caution. The important message is that harvesting water during a very dry year is more likely to have an effect on the estuary, particularly in summer months.

Conclusions

The budget and ecosystem model comparison of the drought years of 2006 and 2007 with the relatively wet years of 2004 and 2005 predicted that the major reduction in river flow would have led to a decrease in the average concentration of nutrients in the estuary, and a subsequent decline in the productivity of phytoplankton, oysters and benthic microalgae. In contrast, the model predicted that the biomass of seagrass will have increased in the drought years due a reduction in senescence caused by epiphytic overgrowth. The key question that follows from a water management perspective is what effect more subtle changes in river flow might have on estuarine dynamics. In this regard, perhaps the most important model prediction is the nonlinear response of the estuary to increasing river flow, leading to greater benefits per ML of river flow at low flow than at high flows.

At low river flows, the benefits to the estuary per ML are far greater because the nutrients spend more time in the estuary due to the longer flushing times, and thus are more likely to be taken up by the algal primary producers. In contrast, as river flow increases there is less time for biological uptake and so benefits are smaller because the flushing time is short. It is important to note that these results are not suggesting that there is no benefit to the estuary at high flows, but simply that the greatest benefit per ML of river flow is at low flows. As it stands, the Water Management Plan has cease-to-take periods for low flows (<7.6 ML/day November to April and <9.5 ML/day May to October) based on the environmental water requirements of the Little Swanport River. However, at the time of developing the Water Management Plan there was very little information available on the environmental water requirements of the estuary. The results of the research reported here certainly support the cease-to-take flows given that the greatest benefits per ML to the estuary were at flows less than 20–30 ML/day. The only point of difference with cease-to-take flows derived for the river is that, from an estuarine perspective, the model predicts that the benefits from river flow are greater in summer than in winter because many of the ecological processes increase with temperature and are therefore more responsive at higher summer temperatures.

We assessed the implications of the increased water allocation. Perhaps not surprisingly there was no discernable effect on the estuary predicted when the allocation was increased by just over 2000 ML/year in a dry and average year when the total flows were between 25 000 ML/year and 68 000 ML/year. However, when we re-ran the model to assess the effects of full allocation in 2007 (when the total flows were 4258 ML for the year) there was a detectable effect most notable in summer. This result should be treated with caution given the uncertainty inherent in model simulations. Nevertheless, the important message is that harvesting water during a very dry year is more likely to have an effect on the estuary in providing valuable ecosystem goods and services, particularly in summer months.

Caveats and future directions

The ecosystem box model was developed as a simple representation of the Little Swanport estuary to simulate the internal nutrient cycling processes among its major primary and secondary producers and the physical exchanges across the ocean boundary. It would be a mistake to think that the model and our understanding of the estuary dynamics are complete. Unfortunately, we were only able to validate the model's performance under low flow conditions, and, as such, when more field data become available under different river flow conditions, this model will be refined further, ensuring greater certainty in predictions.

In this study we have focused on the cycling of nitrogen because it has long been recognised as the limiting nutrient in coastal systems; however, other nutrients such as silicate and phosphate can also be limiting (see Crawford *et al.* 2010 for data and analysis on silicate and phosphate). Although there was no evidence of silicate limitation in the empirical data or in the initial model runs when it was included, our data on silicate loads during floods and concentrations in the estuary are limited, and as such our findings are by no means conclusive. DIN:DIP ratios in the water column confirmed that the system is more likely to be nitrogen limited generally. However, unlike DIN, DIP did not increase in concentration with river flow, indicating that the system may become phosphate limited during floods. A more complete understanding of the potential for nutrient limitation in Little Swanport would be best resolved by experiments involving adding nutrients and detailed sampling during flood periods.

This research highlights the importance of physical exchanges across the ocean boundary, and the critical importance of river flow in dictating the magnitude of this exchange. It also highlights the need for more detailed data on the nitrogen reservoirs in the ocean so that their movement across the estuary–ocean boundary can be predicted more accurately. Greater spatial resolution of the physical exchanges, and therefore ecosystem responses within the estuary, is also likely to improve the accuracy and utility of the model.

In an effort not to make the model too complex, the system was represented as a single well mixed box, and for the purposes of the study we were able to predict system-wide averages reasonably accurately. The presence of both vertical and horizontal (along estuary) salinity gradients during significant river flows illustrated more complex two-dimensional hydrodynamics were present in the estuary, and consequently it is likely that some degree of increased spatial resolution would also improve the future utility of this model.

12

A short history of the catchment settlement

Tor Hundloe and Michelle Wenner

As already noted, our catchment has a human history of some significance – feeding approximately one-tenth of the Indigenous Tasmanian population for millennia. Of this we know little, and so it will remain indefinitely. There are no history books to read. However, it is for our research, at the very least, to sketch the available history of the catchment, because history does matter, and there is invariably what economists call 'path dependence' – what occurs today can be traced to what happened yesterday, the day before and the day before that, going back to some earlier 'tipping point' in history. The catchment could have remained a natural area, with no land clearing, no farming, nothing but nature. If that had been the case we would not have spent three years studying the catchment – at least for the reasons we did. It would have been very interesting to have ventured into it in its pristine state. Of course, no sooner do we write that but we recognise that we cannot exclude the role of the Indigenous people in shaping the landscape and utilising its resources. Humans in some small way have placed their footprint on the catchment for tens of thousands of years.

The first Tasmanians: the Palawa

Anthropologists inform us that Indigenous Tasmanians began to arrive on the island now known as Tasmania between 50 000 to 35 000 years ago. These people, collectively known as *Palawa*, were, by the time Europeans arrived about 200 years ago, a people separated from the rest of humanity for an extremely long period in historical terms. The land bridge that connected Tasmania to mainland Australia submerged some 10 000 to 12 000 years ago as the last major ice age ended and sea levels rose substantially. From that day on the Palawa were by themselves – a few thousand living as hunter-gatherers. No trade, no exchange of genes, no exchange of ideas with other human beings for tens of thousands of years. Consider the fundamental changes to human societies after their separation from mainland Australia. The agricultural revolution occurred, Alexander the Great conquered most of the known world, virtually all the animals useful to humans were domesticated, gunpowder was invented, large cities with sewage and sanitation systems formed (such as ancient Athens and Rome), large ocean-going vessels were constructed, and a multitude of wars, the plague, and other diseases took their toll on humankind for millennia. Maybe the Palawa were lucky people to have escaped much of this.

The Palawa developed into nine separate nations spread across the island. Mulvaney and Kamminga (1999) suggest that there were many more smaller divisons. Each tribe or nation was obviously not many in number, as the total Tasmanian population has been estimated to be in the low thousands. When the British colonists came they named the Indigenous people the 'Big River Tribe' and the first Indigenous Tasmanians inhabiting the Little Swanport catchment were known as the 'Oyster Bay Tribe'. Popular belief is that the latter tribe was most likely the largest in Tasmania, consisting of between 600 and 800 individuals out of a total of approximately 5000 to 7000 inhabitants (Robson 1983, p. 20).

The territory of the Oyster Bay Tribe extended along the east coast from the northern point of St Patrick's Head south to the Derwent River. Inland, their boundary ran south-west along the Eastern Tiers to St Peter's Pass at York Plains, and reached in a southward trajectory towards the Derwent Estuary at New Norfolk. As evidenced by the numerous large shell middens found on the coast, the main diet of the Oyster Bay Tribe was shellfish (in particular, native oysters). Seafood was supplemented by kangaroo, possum and various root vegetables. During the summer months the people frequented Schouten and Maria islands to collect birds' eggs.

Mulvaney and Kamminga (1999) devote a chapter in their book to the prehistory of Tasmania. The interested reader is directed to it and the substantial writings of Rhys Jones (considered the pioneer of Tasmanian archaeology). All we can do here is summarise a few scant points made by the experts. One of the obvious ones is that, due to their isolation, the Indigenous Tasmanians 'missed out on innovations that became widespread across south-eastern Australia' (Mulvaney and Kamminga 1999, p. 344). Another is that the dingo, which had come to Australia when Aboriginals came from Asia, never got to Tasmania. Bass Strait was formed before the dingo had the opportunity to enter Tasmania. When the British brought their European dogs they were adopted within a generation by the Aboriginals as hunting companions – and, fascinatingly, were incorporated into Indigenous mythology (Mulvaney and Kamminga 1999, p. 344). The third matter worthy of mention is the debate as to whether or not the Aboriginal people ate scaled fish. The conventional wisdom that they did not is due to a traditional cultural taboo in their creation myths (Mulvaney and Kamminga 1999, p. 344). On the other hand, the local people were healthy diners on all other marine life, oysters, various other shellfish, crustaceans, stingrays, and mammals (for example seals), as well as water-dwelling birds such as penguins and mutton birds. Tragically, there is not a great deal more than these few points that we can record. The Indigenous population was decimated in the wars between the European settlers and the Indigenous people that transformed Tasmanian society in the nineteenth century. Those who were not killed were transferred to Flinders Island in Bass Strait. Today, there are numerous descendants of the original people living in Tasmania, a result of considerable intermarriage.

We have mentioned previously ecosystem goods and services in relation to modern, industrialised societies – where the ecosystem, unfortunately, takes a back seat to the products we buy and sell in the marketplace. For the Indigenous Tasmanians, the ecosystem – nature – was their equivalent of our farming and manufacturing combined. Everything they needed for survival was there for the taking. That is not to suggest that life was easy. Hunting and gathering can be arduous, time-consuming work. The one major tool used by these people was fire, for the purpose of driving native animals (in particular kangaroos) from the bush into the open.

While it is possible to describe and analyse what economists call 'primitive economies' (sometimes called 'primitive communist' societies because of their communal nature and the non-existence of property rights in land), we won't attempt this with the Palawa. Their small number and the abundant resources available to them meant that the fundamental requirement for something to have economic value, a scarcity of the required resource, did not exist. Whatever we call their lifestyle, the free ecosystem products were the means of life (food, clothing, shelter, and the source of their myths).

The relations between colonialists and Indigenous Tasmanians cannot be viewed through a single lens of colonisation, subjugation and racism. There was an array of exchanges – the exchange of language, culture, food, dance and song, in addition to periods of mutual violence and hatred. Based on colonial reports, the early days of European arrival and settlement were characterised by relative harmony between the Europeans and the Indigenous Tasmanians. Both groups were small in number. It was only when the Europeans commenced large-scale land settlement – and treating land as private property, including fencing it – that serious conflicts occurred. By then it was too late for the Indigenous people. They were outnumbered and lacked the weapons (the guns) of the Europeans.

The first Europeans to establish either permanent or semi-permanent camps and shelter in Tasmania, the sealers (followed by the whalers), were likely to have local women as their companions or common-law wives. This suggests a degree of harmony between the two peoples. Tragically, however, harmonious relationships eroded with the range of atrocities that were committed by some Europeans, for a variety of self-interested reasons. Bushrangers, whalers, sealers, convicts and officials were all responsible at times. Harmony disappeared completely in the Black Wars from 1828 to 1832.

The Little Swanport area plays a minor but not insignificant role in the relationships between Europeans and the Aboriginal peoples. There are reports of the Aboriginals stealing sheep and burning down houses within the Little Swanport area. Louisa Anne Meredith, wife of the pioneer settler Charles Meredith (a poet and early ecologist), describes these incidents in journal entries that she kept while travelling through the catchment. Meredith presents an account of how one of the local innkeepers, who had been many years in the area, escaped being victim to the 'first murderous attack made here by aborigines on settlers' (Robson 1983, p. 77). There is no concealing the fear that the early settlers of the area held. To what degree these fears were appropriately directed at the Aboriginal people is a moot point. It needs to be borne in mind that sheep stealing and other crimes were common actions of bushrangers and, without firsthand knowledge, who was to blame was at best a guess, at worst an allegation based on prejudice.

Early European explorers

European history begins in Tasmania in 1642 with the discovery of the island by Dutch navigator Abel Jansz Tasman. He named the land in honour of Anthony Van Diemen, Governor of the Dutch East India Company, then established in what is modern-day Jakarta. Among other things, one relevant to our area of interest, Tasman mapped Schouten and Maria islands, both located adjacent to the coast of the Little Swanport catchment.

Captain James Cook arrived in Australia on 19 April 1770, and the reports back to England by his botanist, Joseph Banks, encouraged the British to use Australia as a penal settlement. Soon after Cook's voyage, French explorers, Captain Marion du Fresne and Marques de Castries, are thought to have been the first Europeans to make contact with the Indigenous Tasmanians. An Aboriginal was killed during this contact. In the same year the English captain Tobias Furneau arrived in Tasmania, landing on Bruny Island.

In 1789, Captain Cox is believed to have been the first European to explore Great Oyster Bay and the Swansea area. The passage between mainland Tasmania and Maria Island is named Mercury Passage after the brig that Cox sailed. Cox noted that there were Indigenous inhabitants in the area who subsisted predominantly on oysters and other forms of shellfish. Frenchman Nicholas Baudin, on a scientific expedition in 1802 aboard the vessels *Geographe* and *Naturaliste*, charted the Freycinet Peninsula.

European colonisation and settlement

The so-called First Fleet arrived at Sydney Cove in 1788. It consisted of 11 ships, carrying 1030 people (of which 736 were convicts), plus stores. The loss of the North American colonies after the War of Independence led the British to use the newly 'discovered' Australia as a penal colony for its often unfortunate underclass, guilty of such minor transgressions as stealing a loaf of bread to feed their hungry children. Building prisons in their home country was considered too expensive, and ship hulks acting as prisons moored in the Thames were overflowing. The vast open lands of an island continent from which there was no escape – back home to Britain at least – was the solution for the upper-class rulers of Britain.

Formal colonisation of Tasmania commenced in 1803 in the south, and then in the north, of the island (in what we know of today as Hobart and Launceston, respectively). The original Hobart colony was at Risdon Cove. It comprised only 49 persons, of whom 21 were male convicts and three female convicts. They arrived with Lieutenant John Bowen in September 1803.

From the earliest days the matter of providing meat, leather and fibre was settled. In September 1803, 16 ewes, four ewe lambs and three wether lambs were landed at Risdon Cove – they were crossbred sheep. Kirkpatrick and Bridle (2007, p. 3) report that in 1803 'there were extensive areas of grassland and grassy woodland in Tasmania, all well suited to immediate occupancy by European stock, but also a major source of sustenance for the Indigenous people.' By 1809, a formal count showed that there were 1091 sheep in Tasmania. Already there was some influence on the quality of the wool due to cross-breeding with the Spanish-type merino. In 1820, the Governor, Sorell, bought 300 rams from James Macarthur of Camden (Kirkpatrick and Bridle 2007, p. 8). By 1821, wool production outstripped local demand and the first export of Tasmanian wool had occurred; from 1827 to 1831 the amount of wool exported to Britain increased from 192 075 pounds to 1 359 203 pounds. A sheep census in 1820 showed there were 182 468 sheep, while a year later only 7400 non-Indigenous humans were recorded (Kirkpatrick and Bridle 2007, p. 11) or one-twenty-sixth the number of sheep. Before 1820, not a great deal of land had been assigned to the farmers; however by the 1820s 'most of the Midlands and the central east coast were granted' (Kirkpatrick and Bridle 2007, p. 9). We are getting ahead of the story.

The first Europeans to visit the Midlands area – the western extremity of the Little Swanport catchment is in the Southern Midlands – was a party led by Lieutenant Thomas Laycok. This was in 1807. By 1812 the Tasmanian authorities, under instruction from Governor Macquarie in Sydney, began surveying a land route from Launceston to Hobart Town (as the latter was then known). The number of convicts being sent to Tasmania was increasing and the need for farming and pasture land subsequently increased. Escaped convicts (bushrangers) were practising their trade throughout the Midlands, and the ability to move more troops and provide security for free farmers was becoming a priority. A road between the two towns would be a great benefit to farmers transporting products and troopers showing their presence in the developing pastoral country.

Settlement of any significant scale did not occur within the Little Swanport catchment until the 1820s. By then there was considerable European presence and well established farms and communities in other parts of the Midlands and the start of settlement on the east coast. It was on the coast that the first concentrated farming in the catchment occurred. The original settlers of the coastal area were the Meredith family and the Amos brothers.

Box 12.1: Early life on the land

All the pioneers' first homes were humble dwellings compared with what they had in Great Britain, or on the continent of Europe. The courage of those early pioneers was truly amazing, and it is through them and their striving to make a success that we are enjoying our wonderful island today! Bless them! They were up against fearful odds in their completely new country. Droughts came (there were no droughts at home of course) and their pastures and crops failed, and floods occurred that carried away sheep and soil from ploughed paddocks. Then fires burnt down their barns or homes, and the deadwood fences.

When the Whites came, they took over the virgin land, tilled the soil, grew crops in it for food, and made bricks out of the clay. They cleared the land and turned the large trees into timber for building homes, furniture and ships, and small trees piled up in long rows to form fences between paddocks. They ran sheep and cattle on the pastures or on the 'runs' in the hills. The wool was shorn off sheep's backs and used for spinning or weaving into garments. They used dyes made from the vegetation, such as flowers, leaves, bark and seed pods. The skins of cattle and sheep were tanned into leather. But often raw or untanned bullock hide was used. This would be lightly salted and dried, then hammered to render it more supple and made into reins for horse or bullock teams. It was cut into long, narrow strips three quarters to an inch wide (two to two and a half cm). Raw hide was also used for flexible hinges, soles for boots and it had many other uses.

Skins to be tanned were put into a pit of tannin water, sealed up by government inspectors and left for 12 years! Then the pit was unsealed and the contents allowed to be taken out and used. The leather of the 'old days' was very much better than the modern leather.

Rivers and streams were harnessed for irrigation and to turn water wheels which provided the motive power to grind wheat into flour.

Source: Shaw (2000)

They settled in the area north of the catchment, around Swansea (then known as Great Swanport). This was in 1821. The Talbot, Buxton and Cotton families followed soon after.

To set the scene for what follows, it helps to sketch the trials and tribulations of the pioneers. Freed convict, escaped convict, bushranger or 'rich pastoralist', all faced the same physical and economic environment. Many did exceptionally well. A perspective of life in the early days is presented in Box 12.1.

Settling the lower catchment

Because there are significant differences in how the upper and lower catchment were settled, it is important to tell their stories independently. George Meredith and Adam and John Amos are the names associated with the pioneering history of Swansea and to a lesser extent the area south, the Little Swanport catchment. When Meredith arrived in Tasmania in 1821 he was soon engaged in dealings with the Governor William Sorell. Meredith was a wealthy Englishman who claimed to have descended from Welsh royalty (Nyman 1990, p. 1) and was keen to farm an uninhabited part of the colony. Sorrell agreed to the concept and suggested the east coast. He assigned 2000 acres (809 ha) to Meredith in a location to be determined.

The Amos brothers and their families were on the same ship to Australia as Meredith. The Amos family were Scottish farmers. They were not rich, and when the opportunity arose to enter into a tenancy arrangement with George Meredith they did. The latter had convinced them of the potential benefits to be had in the new colony. Meredith was not acting purely in friendship. John Amos had virtually no money or possessions (Adam was better off) and George Meredith had with John Amos the perfect means of acquiring more than one person's share of the colonial land that was being given away. Meredith drew up a legal document which required John Amos to surrender any deeds to land granted to him by the Tasmanian Governor. In return Meredith gave John Amos the rights to manage 300 acres (121 ha) for eight years, after which Amos would obtain as his own 100 acres of the 300; the remainder would go to Meredith (Nyman 1990, pp. 9–10). Such was the laissez-faire capitalism of the early colonial Australia. When Meredith obtained his 2000 acres, Adam Amos got 500 and John Amos 400. The appropriate area for settlement, designated to be on the east coast, had to be found.

A party set off in a whale boat, travelling up the east coast from Hobart. On their way north, Adam Amos, with the assistance of a local shepherd, was taken to the Little Swanport River. The party discovered that living off the land – as they undertook their search for the most desirable land for farming – was easy. George Meredith, using skills he had acquired in bird-shooting back in England, was able to shoot in one evening two ducks, a pigeon and two cockatoos. The party explored the Little Swanport River environs; they were attracted by the beauty but disappointed by the quality of its soils (Nyman 1990, p. 13). Consequently, they kept heading north until they reached Oyster Bay, which the magnificent Freycinet Peninsula protected from the waters of the south Pacific Ocean. Nyman (1990, p. 15) informs us that 'here was a large expanse of virgin land ready for the taking – an ideal place in which to start the settlement of his [Meredith's] dreams'. And so Meredith and the Amos brothers took up their land grants at Great Swanport (today's Swansea). As this area is well outside our catchment we shall not dwell further on its history, but our interest is in the manner of how land was obtained. The laissez-faire process explains how fortunes were made in the early days of colonisation.

There is reason to discuss the arrival of Thomas Buxton in the lower catchment. Buxton commenced his career as an east-coast farmer working as a farm manager in the Great Swanport area. In mid 1823 Buxton applied for 500 acres (200 ha), approximately 19 km south of today's Swansea. Being granted the land, he named it Mayfield (after Mayfield House, his Derbyshire home). The southern boundary of the farm is Buxton River, which we have come across in Chapter 8. The grazing property of Mayfield is within the boundaries of the 'extended' Little Swanport catchment used for planning purposes by the Tasmanian Government, but it is not part of the geographically defined catchment.

The so-called Black Wars did not pass by the east coast. In late 1823, a dispute erupted between a large group of Aborigines[1] and settlers at Grindstone Bay, considerably south of the Little Swanport River estuary. Among the Aborigines was Muskitoo (who had come from Sydney and was one of the better known Aboriginal fighters). Little Swanport was not a settlement like Great Swanport. There were only huts at the former where shepherds and stock sheltered, but little else. Both Aborigines and bushrangers were the threats to the isolated shepherds and farmers. Bushrangers were a constant concern, and shepherds' huts around Little Swanport were easier prey than the more settled areas; for example in the mid 1820s a shepherd's hut was ransacked (Nyman 1990, p. 41).

There were probably hundreds of bushrangers living in the Tasmanian countryside in this period, none more infamous than the Brady gang. Brady was captured in the north of the colony in 1826 and hanged soon afterwards. As the number of convicts to the colony increased – and as the bushranging problem continued – the Colonial Government sought a new penal settlement. Maria Island on the mid east coast, south-east of the Little Swanport estuary, was the chosen site. The first prisoners were taken there in 1825. If this suggests a lawless society, this was not the case. It is always the few who establish an undeserved notoriety for the rest. Keep in mind that in this same era there was a far greater degree of lawlessness in the United States – and its lawnessess extended much closer to the present (some argue that it still exists). Famous names such as Wyatt Earp, Doc Holiday, Wild Bill Hickok, Billy the Kid, Jesse James, and the two sharp-shooting ladies Pearl Hart and Belle Star, made their names in the latter part of the nineteenth century in the 'wild west' of the United States. By this time Tasmania was settling down to the hard grind of peaceful farming. Judge Tasmania fairly.

Of all the locations that today are part of our 'extended' catchment, Mayfield was considered a dangerous place, being somewhat isolated from the larger settlements to its north and south. There was a major Aboriginal attack on Mayfield in March 1826. Nyman (1990, p. 55) states that there were 80 to 100 Aborigines involved. Property was stolen and the winter supply of hay was burnt. Several Aborigines were killed.

Settling the upper catchment

Governor Macquarie travelled through the Midlands in 1811. Ten years later he named the site of present-day Oatlands – 'a land for oats'. According to Macquarie it was 'well watered' and with 'rich fertile soil', reminding him of his native Scotland where oats was the major crop. In 1825, a military detachment station was opened in the yet-to-be-built town. The troops were sent to protect the farmers and the skilled tradesmen who were beginning to settle and work in the locality. A significant number of tradesmen were consigned to establish the foundations of the town and they either were under threat, or felt threatened, by the local Aborigines. Notwithstanding that, priorities had to be sorted. By 1829, a brewery was under construction. More breweries followed. It might have taken nearly 50 years, but by 1877 the town had seven hotels, three breweries and an aerated-water production facility. Tasmania is known to this day for its excellent beer.

During the 1830s much building of houses and civic buildings took place in Oatlands. Many remain today. Convict labour was plentiful, and this was what made possible the grandeur of many buildings. Keep in mind that until the penal system was abolished, Tasmania (as with most of European-settled Australia) was, in part, a slave economy. Not only did the Colonial Government use convicts on major infrastructure projects (such as road building), but the free settlers, particularly those from the larger grazing properties, benefited from the free labour of assigned convicts. The convict-built buildings of Oatlands remain today as a testament to a different form of economy and architecture that existed then. With free labour, much splendour is possible.

As early as 1861 the Municipality of Oatlands existed as a formal local authority, and only in 1993 was it absorbed into the larger Southern Midlands Council.

There is a part of the upper catchment that, to this day, is a different farming environment from the large grazing properties on the coast and in the north and east of the Southern Midlands. Before European settlement the upper catchment was a mix of natural grasslands and timbered hills. A few remnant timbered hills have survived, and natural grasses are still a major feature of 'run country' for sheep. The grasslands covered much of Southern Midlands and were perfect for sheep grazing. Freed convicts, escaped convicts, bushrangers and eventually free settlers moved into the Midlands early on because of its grazing potential and its native wildlife as a source of free food, and for the escaped convicts and bushrangers, shelter and hide-outs provided by the timbered hills. No wonder the area was settled quickly – it had something for everyone. Until the land grants, the locality was a virtual 'commons'; unfenced land where shepherds attempted to manage small flocks, always alert to theft from other settlers out to establish farms or simply feed themselves if on the run from the law.

The area that today is conventionally associated with the upper catchment of the Little Swanport is the Woodsdale area. Here there were not vast grassed plains but rather hills, dales, valleys, and significant patches of timber. There were, naturally, Aboriginal people in the area, and the first Europeans to this area were freed or escaped convicts. It was tough country, but for people who found themselves 'free' in a resource-rich environment, it was a near utopia.

The first settler of the Woodsdale area was Martin Bourke, who arrived in 1845, taking up land at the spot then known as 'The Bluff'. This was followed soon after by the Palmers in 1869, who settled at Tin Pot Marsh. Within the next decade the Scott, Rowlands, Gerlach, Brooks, Wagner, Howard, Johnson, Wiggins, Crawford, Smith, Hayers, Roberts, Powell, Kenna, Byers and Kingston families moved into the Woodsdale area. Today most people living in this part of the catchment can trace their family back to one or more of these families (Dare 1989). The links with the past (based on the intermarriages between many of these families) are today the strong glue in cementing the 'bonding' social capital that exists in the upper catchment. This will be discussed when we come to our research findings, but note here that what we witness today in this tiny part of Australia is the type of close community destroyed elsewhere by population growth and agribusiness (as opposed to family farming). In this part of our catchment we have, in all probability, one of the last stands of the small-scale independent farmer in Australia. Much farming in Australia, as in North America, has become large-scale agribusiness, yet it remains family-based and small-scale in many parts of Europe, Asia and Africa. We are digressing too far, but we believe it is important to draw attention to the type of farming remaining in our catchment.

The early farmers were a fascinating mix of poor, free settlers (although many were married to pardoned convicts) or the sons and daughters of convicts, wandering bushmen (shepherds, kangaroo hunters and the like) who finally settled in one place, and wealthier land 'grantees' from the United Kingdom. All had to live together, and life was tough and good. There was no advantage in being a 'toff'. This was a classless Australia – the foundation of a strong community that exists to this day.

Notwithstanding the hardships of living in tiny log cabins, sunrise-to-sunset back-breaking work in clearing 'scrub', tending crops and gardens, and looking after large numbers of children, life was generally better in Tasmania than in the home country, the United Kingdom. This was obviously the case for the freed convicts, and more often than not for the poorer free settlers. Nature provided abundant wild game (kangaroos, possums, echidnas, wombats, and numerous birds). These were the ecosystem goods of our great-great grandparents' era. No one went hungry as they were likely to in the United Kingdom. The 'bush food' was there to be taken by these European 'hunters and settlers'. Had environmental economists been asked to estimate the value of nature's goods in the 1800s, the monetary value they are likely to have come up with would have been very large. It is likely to have exceeded the monetary value of marketed goods and services in that era. One did not go to a supermarket for food and other household needs, nor to a clothing shop for anything but 'the Sunday best' attire. A rifle or trap provided an extensive range of game meats. An old wether provided all possible mutton cuts. Flour and sugar bags, once emptied of produce, became the materials for bedding and even clothes, supplemented by treated animal skins. The story is different today. In our study of people living in the Little

Swanport catchment in the twenty-first century we enquired as to the benefits they obtain from the environment at no cost. Only firewood and fish were mentioned. The past is forgotten except for a few that make the time to record family histories. Without their commitment we would know little of our catchment 100 years ago. Even with their efforts to draw on, we are likely to know more about life in Britain in Shakespeare's time.

Doreen Dare, who compiled *A History of Woodsdale* (Dare 1989; updated by her daughter Julie Bellete: Dare and Bellette nd), informs us that the first settlers arrived in Woodsdale in 1875. However, many had bought their land before that, in the 1860s. It was, in contrast to the grasslands of other parts of the Southern Midlands, a heavily timbered area, and a great deal of hard manual work was involved in clearing it for farming. The farms were, and are, small. The area was, and is, closely settled. By the First World War it had a population of 184.

Doreen Dare with Julie Bellete compiled brief stories of the early settlers of Woodsdale. Doreen Dare writes in her introduction:

> *some Woodsdale ... people descended directly from the expedition Lieutenant Collins brought to Van Diemens Land in 1803. Some can trace their ancestry back to the marines and some to the convicts ... some can go back to the first settlement to Sydney in 1788, and (some)who went to Norfolk Island and later to Van Diemens Land'. (Dare and Bellette n.d., p. 81)*

Doreen Dare asks us to imagine life in the very early day:

> *Can you imagine what it was like for those convict settlers. Many were not able to write and were sent against their will to this distant land and probably never had contact with their family again ... Life would have been very hard for the women; many were married by the time they were 15 and went on to rear large families. There were no doctors closer than Oatlands, and to get to a doctor one either had to walk, ride a horse, or go in a horse-drawn cart. A bond was formed between the residents who helped each other in times of sickness and bereavement. It became a close knit community of helping one's neighbour in times of trouble and in times of joyful celebration. (Dare and Bellette n.d., p. i)*

Dare and Bellette also inform us that the early settlers had a vast range of backgrounds. In the early twenty-first century as we went about our research, meeting and talking with dozens of farmers, their wives and families, we found the same attitudes, and the same community spirit, if a little more subdued as twenty-first century attitudes to privacy prevailed. Notwithstanding the popularity in sharing trivia of a personal kind on Facebook and the like, people tend to be desirous of privacy. We tend to expect too much rationality from humans.

The fact that the early settlers had large families, that considerable intermarriage occurred and people tended to stay rooted to the district, means that the names of these people are still common today. Doreen Dare gives the following as an example: Wiggins, Wagner, Palmer, Rowlands, Byers, Kingston, Howard, Scott, Crawford, Roberts and Kline. Of course, today many descendants are spread far and wide. This is inevitable in a small farming community with limited land. One of the better known descendants is Simon Wiggins, who joined the Melbourne AFL team Carlton as a player in 2001. Thomas Wiggins was the first child to be registered in the colony of Tasmania, and that was in 1804. His father was a Royal Marine who brought his wife and daughter to the new colony where Thomas was born.

The large families of the settlement era did not mean that all went on to live long lives. Doreen Dare writes that most families were of 10 or more persons, and they were lucky if all children survived. It was not only difficult for sick children far removed from medical help to survive; occasionally women died in childbirth. While nowhere near as life-threatening but certainly important, the other health problem was dental hygiene. Considerable effort was necessary to visit a dentist. There was a simple solution, as Doreen Dare informs us: usually all the teeth were pulled out at one or two sittings.

In this heavily timbered area of the upper catchment, the main technique in clearing forests of the larger trees was to ring-bark them. When the trees died (after some considerable time) the dead timber was burnt. Saplings and bushes (or 'scrub') were cut down by axe and burnt. The stumps had to be removed. The larger ones were dragged out by bullock teams. Small and difficult-to-remove stumps were 'grubbed' around with picks and mattocks before being pulled out. In the more recent era, starting after the Second World War, the land was cleared by bulldozers – months of manual labour became days of mechanised labour fuelled by petrol or diesel. Before the dozers, the area we are discussing was a patchwork of paddocks 'here and there through the bushland', today it is 'the occasional piece of bushland amongst the paddocks' (Dare 1989, p. 4). Dare does not discuss the impacts of settlement other than this. The change in microclimates, and the effect of significant loss of tree cover on rainfall evaporation and run-off have, presumably, been experienced by subsequent farmers. The environment is different today, yet many of the people of the upper catchment have communal values and attitudes not that different from those of their great-grandparents. Modernity is put into context when one is invited into communities that are both modern (all the information technology available) but conservative in the true sense of caring for the good things of the past.

We will continue the story. Farming in the early days was largely a subsistence activity. The first task was to produce enough food for the family (see Box 12.2). Milk, meat, vegetables and fruits were the necessities of life. To produce milk and animal protein in highly variable climate with relatively cold winters required growing feed for the cows. However, some things came before this. The first priority following the construction of a very basic house (what the Americans would call a log cabin) was the preparation of a vegetable garden followed by a fruit garden. Berries, as quick-growing fruit, were popular (gooseberries, strawberries, raspberries), as were redcurrants and blackcurrants, cherries and plums. Apple and pear trees were planted, but years passed by before they bore fruit. Once the garden was planted, a poultry run was established with the help of neighbours who supplied eggs for setting, as well as roosters to do their job. As the human population grew, particularly in the nearby townships, eggs were sold and became a minor source of income for the farmers, as did the sale of poultry. Geese and ducks brought good prices at Christmas and Easter.

Once the farms were established, the major farm income was earned from milking dairy cows for home-made butter production to be sold locally or sent to Hobart. When the butter factory opened in Hobart, farmers began to send cream to it rather than 'beat' the butter on the farm. However, there was also local production of butter. At the end of the nineteenth century one of the local upper catchment farmers bought a cream separator and established a small, local butter factory. By the early twentieth century, a carrier engaged in transporting products (such as cream and eggs) to Hobart had commenced operation, and on the return trip brought grocery items (such as flour, sugar, corned beef, sausages and fresh fruit).

The few cows a farmer owned were milked by hand, which made the 365-day-a-year task a family one, with the children doing the 'stripping' (getting the last drops of milk out

Box 12.2: Breakfast

Breakfast for the man-on-the-land consisted of a large bowl of thick rolled oats made in a double boiler, put on to cook the night before, then boiled up again in the morning. This was followed by bacon and eggs and slices of bread or toast, with butter and honey. Marmalade was not available in those days. If butter was in short supply, dripping was used, spiced with salt and pepper. For a change, cold meat would be eaten. Meat was often eaten three times a day in the country, as often vegetables were scarce. It was too much; boils were a common complaint. These days it is a rare occurrence to have a boil or carbuncle because vegetables are in abundant supply.

Lamb was seldom eaten because lambs grew up to be sheep, and were therefore too expensive. 'Why kill a lamb when the same animal would be twice the size in a year or two?' Mutton was the meat of old sheep.

Source: Shaw (2000)

of the teats). Milking took place before and after school to facilitate the participation by school-age children. During the winter the cows were fed oats.[2] Paddocks had to be ploughed by horse-drawn ploughs, the seeds sowed, and the grain harvested. The horses and the bullocks, if farmers had a team, also had to be fed in winter, which meant that enough oats had to be grown to feed them as well as the dairy herd. Oats were harvested by hand in late summer, using scythes and reaping hooks. Neighbourly help – the social capital of small rural communities – made harvest time something special, with a carnival-type atmosphere.

Being small farms, most were quite close, 2–3 km apart, and large groups of neighbours gathered at one farm this week, another farm next week, and so on until all the harvesting was completed. And the carnival atmosphere moved with the harvesters.

Typical of small-scale dairying anywhere in the world, the farm would have a small number of pigs. They were fed the separated milk. Crops such as peas were grown to complement the pigs' milk diet, and poor-grade potatoes were also fed to them. These had to be harvested, and that was hard work. As winter approached, the chaff had to be cut. First the grain was threshed. This was done by a relatively large steam-driven piece of machinery. Another essential machine was the chaff cutter. Bullock teams were needed to move these machines from farm to farm, they were so heavy.[3] Once the seed had been threshed from the straw (and the fly chaff sieved from the seeds), the seeds were graded, the bigger ones to be used for replanting next season; the pigs got the middle-sized ones and finally the laying hens got the smallest. Woodsdale farming could best be described as mixed farming: a dairy herd, some pigs, some egg-laying poultry, and increasingly potato growing. However, change was coming.

Most of the farmers in the district fattened two or three pigs to make into bacon for their own family's use during the winter. Always one ham was kept for Christmas and another for Easter. They were beautifully cured, kept really well, and tasted so much nicer than our present day bacon or ham. Those were the days when nothing was wasted and I can remember my mother cleaning the entrails to make sausage casings and those which were not used were fried and called chitterlings. This apparently was a great delicacy, but I myself could not bring myself to eat them. Black puddings were made from the blood from the baconers, mixed with breadcrumbs, spices and herbs.

> *The caul fat from the pigs was rendered down into lard and was an excellent shortening for pastry, biscuits and cakes. (Dare 1989, p. 14)*

It was well into the twentieth century, just before the Second World War, that farming changed rather dramatically, not only in the Woodsdale area but throughout the catchment and, we should note, the industrialised world. Tractors replaced horses and bullocks, and petrol-driven machinery replaced steam in threshing and chaff cutting. Not only did the nature of mixed farming change, so did the focus of the farmers. As wartime needs had to be met, farmers were requested to concentrate on milk production. This was an Australia-wide request by the Australian Government. Farming became an essential part of the nation's war effort.

In the Woodsdale area, as elsewhere, dairy herds were increased (if possible), and new dairies and cowsheds were built. The milk was collected daily as transport trucks plied the dirt roads of the farming districts. To provide more cow feed, green crops were planted for the cows to graze. Having done their bit for the war effort, with the war over, the Woodsdale farmers continued on for near a generation with the style of farming they had become used to. This was not to last.

In the middle of the 1960s – again in the small-scale dairying districts across Australia, not only in Woodsdale – the dairy industry changed radically. Milk runs were discontinued except in the major dairy districts (where farms and herds were large). These were, and are, areas with a concentration of large dairy farms and hence there are economies of scale in milk collection. In the Woodsdale area, as with other similar locations where dairies were relatively small and not all the farm was devoted to dairy cattle, farmers converted to beef production. Unlike much other dairy county, the Woodsdale area was also suitable for sheep farming, for both wool and fat lambs, and hence a cattle and sheep mix became common. Wool rather than meat production has dominated, with only small numbers of fat lambs sold. This makes economic sense because we are writing about some of the best wool country in the world – when the rains are adequate! As we write this, fat lamb prices have increased dramatically and another change is underway for sheep-farming communities. The major cash crop in the area became potatoes and remains so today.

Land grants and land uses

Having sketched a little of the fascinating history of the early European settlement, it is appropriate to discuss in some detail the social and economic environment which underpinned the settlement of our river basin. Keep in mind that history matters. The laws regulating settlement in Tasmania closely followed those that existed in the colony of New South Wales at that time. Free settlers were not the only individuals entitled to land grants. Governor Phillips' instructions of 1787 made provision for male convicts who had completed their sentences to be granted land. Although the reliability of records may be questionable (admittedly as with any historical record),[4] in the order of 1500 people were granted land in Tasmania between 1805 and 1823. That is not a lot of people, but there were not many people in the colony.

The peak of alienation of Crown land was reached in 1823 with the grant of 174 000 ha (Kirkpatrick and Bridle 2007). This transfer of land clearly played a significant role in inciting the Black Wars as land granted to be enclosed was no longer a commons for hunting and camping, it was for sheep and farmed crops. Alienation of land in Tasmania came with the condition that all timber remained the property of the Crown, a self-interested decision by Britain at the time, as it had exhausted its timber resources, much needed for ship building. Although land clearing was seen as a necessity for the establishment of improved pasture, it was the existing native grasses that initially supported and made sheep farming viable within the catchment. Today, natural pastures (run country) are important in the catchment. Merino sheep in significant numbers were introduced to Tasmania in 1820. The first export of Tasmanian wool to London occurred in 1821.

The very success of sheep grazing, and indeed grazing in general, is owed primarily to the presence of suitable native grasses. While it is recognised today that fine wool is generally a product of improved pastures, wool growing relies on native pastures to a greater degree than other agricultural pursuits (Kirkpatrick and Bridle 2007, p. 46), and the existence of vast areas of native pasture was a great benefit in the early years of grazing in the Little Swanport catchment, and Tasmania in general.

There is, so it seems, always a downside. The introduction of sheep to Tasmania represented a serious threat to the survival of the thylacine (the Tasmanian tiger). Tasmania is best known for its friendly people, World Heritage temperate rainforests and two animals, the Tasmanian devil and the Tasmanian tiger. From the onset, the thylacine was seen as a danger to stock and was held responsible for considerable loss of sheep. Private and official bounties, along with trappings for zoos and export, and over time the drastic alteration of habitat, led to the thylacine's extinction by 1936. It had no chance. Its reputation was that of an outlaw bent on stealing from the farmers their very livelihood. As with many outlaws, it is unlikely to have deserved its reputation.

The Buckland and Spring Bay Extermination Association was formed in 1884. Its name suggests its purpose, although it did not specifically identify the Tasmanian tiger. The Spring Bay Council placed a bounty of one pound on the head of all adult thylacines. The farmers and the officials got it wrong. The most serious threat to farmers' animals in the early days of farming was not a native animal, not the Tasmanian tiger, and there were no dingoes, only imported kangaroo dogs. Humans were the threat. Due to proliferation of sheep-stealing it was a common expression that 'the only dog to be feared is the two-legged one'. The other threat was drought, and to this very day it remains the greatest threat and one beyond control.

The first farms in the catchment were relatively isolated and their major products were wool and grains while there was, to a lesser extent, fruit growing in orchards, hops production and beef cattle grazing. Sheep hides were a valuable by-product of the wool industry.

> *In the time when there was hardly enough feed for the animals ... [the farmers] cut the gorse from the hedge-rows which they had planted around some of the paddocks along the dead-wood fences. They used a most special cruncher that had a number of flat saucer-like discs with teeth around the rim, similar to the escapement wheel on a clock. In the centre was a raised portion around a square hole for them to fit on a shaft. Another lot of discs fitted on to a shaft between these, giving a very good crunching method, which left no hard spikes to worry the animals, who ate it with great gusto. All the young gorse bushes about the farm were cropped closely before they grew larger with coarse spikes which the animals couldn't eat. (Shaw 2000, p. 17)*

There was a natural side to the rural economy, an ecosystem product. Rural Tasmania in the very early days has been described as a 'kangaroo economy' (Boyce 2006). The kangaroo was 'free', a natural resource available for human consumption at the cost of a bullet, some dog feed, and time hunting. As noted previously, the Aborigines had hunted kangaroos for millennia by 'fire-stick' farming. The new settlers, many extremely poor in their homeland and not used to rations of fresh meat, discovered that kangaroo meat was both tasty and plentiful. This was approaching utopia for a freed convict – or an escaped one. Kangaroo skins, which had provided clothing for Aborigines in the cold Tasmanian climate, became a source for various items of clothing for Europeans. As the sheep flocks grew, sheep skins displaced kangaroo skins for clothing items.

The population of the catchment varied throughout its early history as roaming shepherds, bushrangers and Aborigines utilised the area for their respective benefits, but as free settlers and 'grantees' came to control and own the land the population settled down to a relatively constant number. It was never large. As noted previously, during the First World War the population was believed to be in the order of approximately 180 (Dare 1989). Today it is 521. This is, in population numbers, a tiny catchment, but in area not insignificant.

Places of historical interest

Give the European history of our catchment and the fact that much that was built remains today, we feel obliged to spend a little time on places of historical interest. There is more to this than simple story-telling; the people of the catchment are very proud of their history and its artifacts. Australians struggle with Australian history because those of us of European (or non-Indigenous origins) can look back only so far in their land. Our relatives who stayed in the home countries live with history.

We describe a number of places of interest either in or adjacent to our catchment. How and why they developed is an integral part of our story. Great Swan Port (Swansea) was established primarily as a military garrison. The garrison was withdrawn in 1850, after which the town became the centre of the first police district and first rural municipality of Tasmania. The latter suggests its importance in the early hurly-burly of Tasmanian history. A convict probation station was established at Rocky Hill, just outside the northern perimeter of the catchment. This convict station accommodated up to 400 convicts and supplied convict labour to the surrounding areas. This was a large free-of-cost labour force for graziers and other free settlers who needed work done. A convict-built coach road was constructed in 1834, linking Swansea to Hobart in the south and Avoca in the north. Prior to the construction of this road, the journey was covered either by sea or, for those without a sea-going constitution, by land, which involved a combination of ferry (across the Derwent), carriage where the landscape was kind, and foot where the landscape was not. The overland trek generally took five days.

A construction of note within the catchment is the Lisdillon Salt Works. The name Saltworks remains today. The salt works were constructed using convict labour in the 1830s, on a 2560 ha parcel of land that was granted to James Radcliff, who had come from Belfast. Along with the salt works, a boiling house, windmill, piggery, dwelling house, store, blacksmith's shop, men's hut and overseer's cottage were also constructed. This was some establishment. However, salt production was not a commercial success and it ceased operation in 1841. There is conjecture over reasons for the failure – but it is likely to have been a combination of withdrawal of free convict labour in 1840 along with imposition of import duty on salt. In this case we can't blame the vagaries of nature or the markets, but rather the vagaries of politics.

The Lisdillon property was purchased by John Mitchell in 1854, and by this time the Lisdillon Estate had expanded to 2800 ha. Mitchell was a surveyor from England who had brought eight or nine families from England to work as tenant farmers on the property, where he lived with his wife. This is how the 'landed aristocracy' took its place in the new colony –bringing in labourers if convicts were not available or worthy enough. The pastoral economy worked this way. One had to be rich and important enough to be granted a sizable property and then able to 'sponsor' tenant farmers. It was neither the feudal system nor a classless farming society but a form of 'landed aristocracy' formed around class. The Lisdillon property was so large that a store, church, school-room and post office were built to support the tenant families. Soon after the acquisition of Lisdillon, Mitchell purchased Mayfield and other lesser known properties in the area. His was becoming a virtual farming empire.

Sheep were grazed on the property; crops of beans and peas were grown at the salt works. But times were changing. The property was leased by James Dodge from 1873–1900. John Mitchell died in 1880, and the property passed to his son Mark. The property was sold in 1920 to Sir Henry Jones. Jones's son resided there with his family for many years, until it was brought in 1950s by John Hood, and later by the Cotton family, who came from nearby Kelvedon. The original Cotton family was one of the first to settle in the lower catchment. The property was subsequently divided and sold. It was returned to the Crown in 1983 and is now managed as a coastal reserve.

13

The people and use of natural resources

Tor Hundloe, Peter Daniels, Amy White and Christine Crawford

With the brief history behind us, we now start to develop our understanding of the Little Swanport catchment as it is today. While considerable time will be spent on description, never far from our mind will be the central issues and questions we have dealt with in the lead-up to our case study. If this river basin was a nation (obviously a tiny one, in the San Marino league), what would we make of its economy and society? As a river basin, could its economy and society be improved by changing its present use of water, the resource that defines a catchment? We shall see.

The adjacent towns

There are three small towns on the periphery of the catchment and these play a key role as essential service centres. As will be shown, they rely on the catchment residents for a considerable amount of their income. While not in the catchment, they are of it. Each is introduced briefly and discussed in greater detail later.

Near the source of the Little Swanport River in the north-west of the upper catchment is the town of Oatlands. This is the administrative centre for the Southern Midlands Council. Oatlands draws its businesses from a much greater area than the upper part of the Little Swanport catchment, as it is the centre of the vast Midlands grazing district. Its historical buildings as well as its location make for considerable tourist traffic.

Clockwise from Oatlands, the next town is Swansea, which is across the sheep and cattle plains and down the forested mountains where the sea is reached on the east coast. Swansea is north of the catchment boundary; however, it is a service centre for the north-coast of the catchment. The town is the administrative centre of the local authority. It is also a tourist town. There are magnificent views across Great Oyster Bay to the distant Freycinet Peninsula, a place name that tells us that French explorers passed this way a long time ago – before European settlement. In 2007, Swansea won the prestigious national Tidy Towns award, run by the long-standing environmental organisation Keep Australia Beautiful. Tourism expenditure means that Swansea would certainly exist without patronage from Little Swanport catchment residents. However, that is not to diminish the importance of their expenditure in the town, as we will see later.

On the coast going south are the twin towns of Triabunna and Orford. Both of these are holiday towns for (mainly) Hobart residents. While business from catchment residents is important, they (like Swansea) would exist without it. On the same road, heading towards the capital city of Hobart, in the south-west is the tiny town of Buckland. It comprises a shop-cum-service-station, a hotel, a town hall, a handful of homes, and little more. Before modern car travel it played a greater role for catchment residents than it does today. A day's ride on a horse determined the distance between towns – and each was important, as food, shelter and entertainment had to be provided to the traveller.

Triabunna is the most significant centre on the south-east periphery of the catchment, and destined to become even more important as a holiday-cum-coastal-rural retreat for Hobart residents and retirees. It is a port for the local commercial fishing industry, and looking east one sees great heaps of woodchips pile up, to be loaded aboard ships bound for Japan. There is daily truck traffic as the timber is brought from the surrounding hills – very little of it, if any these days, from within our catchment. Large-scale residential development is on the drawing board and the future is 'upbeat' – but not necessarily for those who came to escape city life. Triabunna and Orford, being on the route to the eastern (lower) Little Swanport catchment, are important service centres for the people who maintain houses in the catchment as weekend or holiday homes, as well as for lower catchment farmers. We will come to describe in due course these towns in more detail, including their economic importance.

The catchment landscapes

The catchment can be divided into three broad-scale landscapes: the upper catchment, which is a plateau with the occasional hill; interrupting the undulating sheep and cattle plains is the mid catchment, which is primarily forested, mountainous and descends rather steeply to the lower catchment, with its hills that gradually flatten out to a relatively flat, low-lying agricultural land with a scattering of hobby-farms among larger pastoral properties and, on the coast, residential/ holiday settlements known as Pontypool and Saltworks Road. Through these landscapes flows the Little Swanport River and its many tributaries, entering into a large estuary that flows into – and interchanges water with – the ocean at Great Oyster Bay via a narrow break. Figure 13.1 shows the land use types over of the catchment today.

The land uses

The major land uses in the catchment are mapped in Figure 13.1. Rainfall, topography, soils and accessibility have determined the land uses across the catchment. All are important, but occasionally one natural feature plays havoc with human aspirations. It is the most changeable, and it is rainfall. Rainfall is highly variable, with a mean of 600 mm per year. This average tells us little, other than it is rather meager. It is neither uniform across the catchment nor reliable enough for short-term planning. Over the long-term one could, if one would, plan around the expected rainfall; however uncertainty would keep one awake at night.

The catchment is covered by the range 400 mm to 900 mm, and much lower rainfall and much higher have been recorded in specific locations in the catchment. As an example of rainfall variability, we have graphed annual rainfall (as recorded at Orford on the south-east coast, close enough to the catchment to be of interest) over a 70-year period. As Figure 13.2 shows, there is dramatic variability in rainfall. This proves a major problem for farmers.

We shall return to analyse the consequences of rainfall variability. It proved to be a major analytical tool in unravelling the value of water allocations to different industries in the catchment. We can note here that stock watering is done via digging and then filling small water-holes (farm dams). There are 1300 in the catchment, with 1200 on farm land. Stock can also access water from the rivulets and the main river.

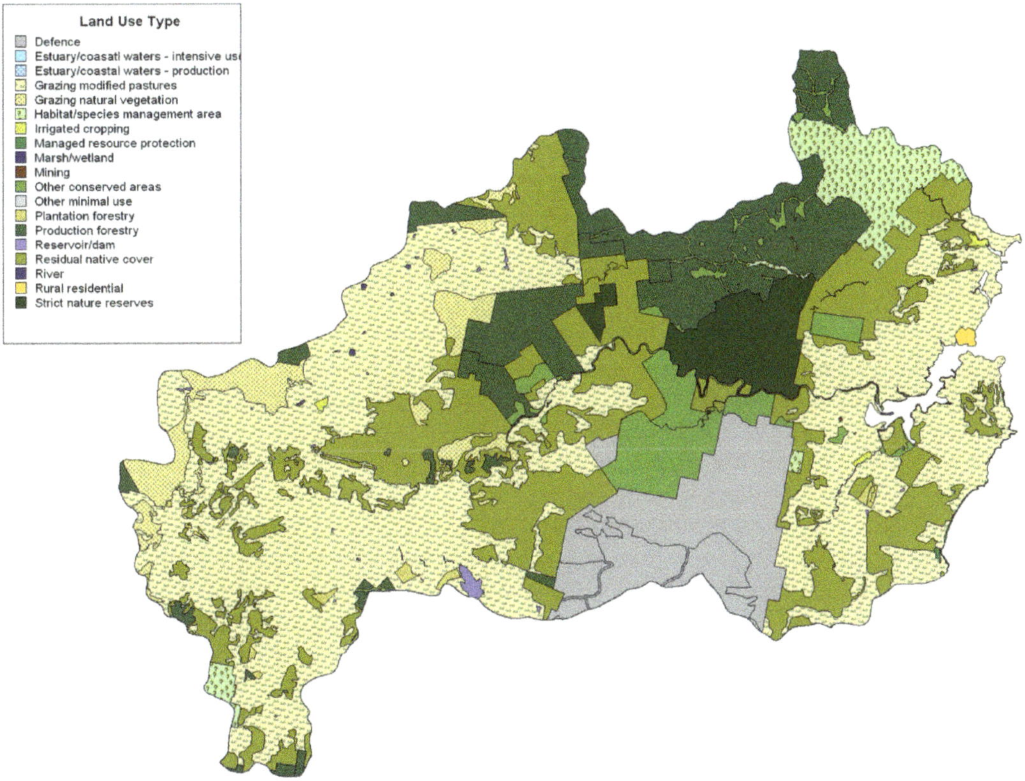

Figure 13.1 Land use map.

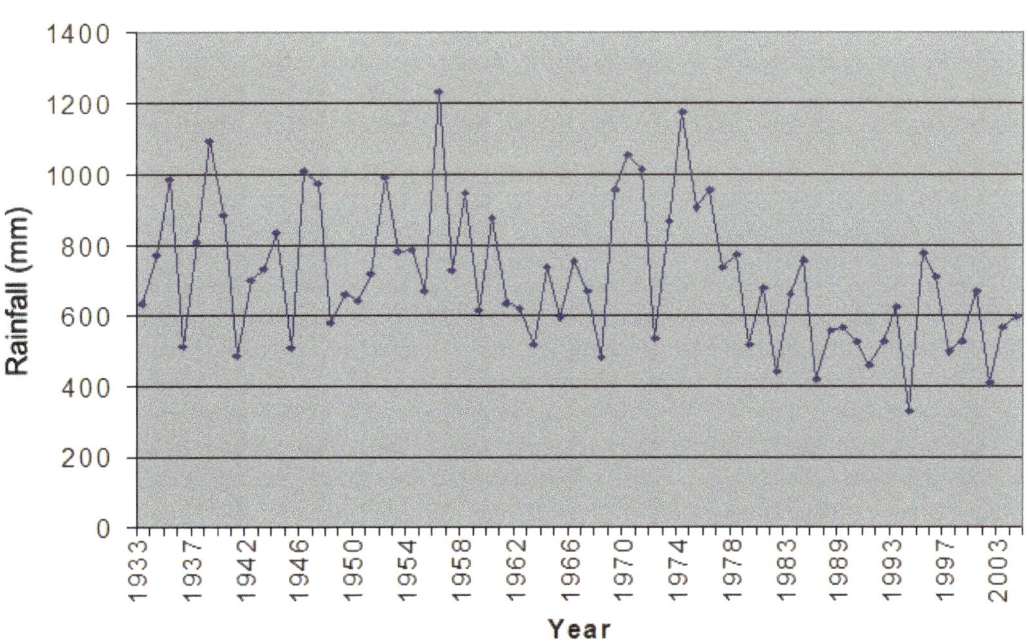

Figure 13.2 Annual rainfall for Orford South.

A real mix of land uses

The cleared land in the upper catchment is a mixture of grazing land (used for both sheep and cattle) and mixed-farming land (that is, crop production plus the running of sheep and/or cattle). Potatoes are the major commercial crop. There are two small-scale tourist operations based on farm experiences; however, they are minor contributors to the regional economy. At the time of writing there is limited forestry (called production forestry) and far less plantation forestry in the upper catchment. This could expand as the large forestry company in Tasmania continues to seek land for native forest plantations. Harvesting existing native forests is extremely conflict-ridden in Tasmania and only plantation forestry is relatively trouble-free, and then only if it is on otherwise 'degraded' land. Grazing farm animals and growing foodstuffs are much more highly regarded – by most Tasmanians – than timber-getting and paper making.

A detailed categorisation of land use types in the Little Swanport Catchment is provided in Table 13.1.

The mid catchment is a combination of old-growth and new-growth forest. Given the propensity of early settlers to search out and sell the most valuable trees, it is unlikely that any of the land is completely untouched. Old growth relates to native forests that have been logged once on a selective basis then left to regenerate to become mature forests, while new growth describes forests that are in the early stages of regeneration. A considerable amount of the forested country is under the control of the Defence Department, and most of the remainder is Forestry Reserve land. There is also forest on private land, some of which is maintained as private conservation areas. Biological diversity (now commonly referred to as biodiversity) is much higher in the forests than in the surrounding farming land. Given the significant area of forest and relatively untouched land in the steep middle catchment, it goes without saying that biodiversity values are significant in this part of the catchment.

The lower catchment has the greatest variety of land users: relatively large grazing properties (some established in the very early days); hobby farms; residential blocks, two B&B tourist establishments; two small olive groves; and an oyster nursery on the banks of the estuary. Within the estuary are two relatively large-scale oyster farms. The oyster industry is of such significance we describe it in some detail – not as much as we would like, due to the fact that 'commercial-in-confidence' protocols prohibit detailed descriptions when a very few businesses are in the same industry. With that understood, we present the following.

Oyster farming

Oyster farming in Tasmania and South Australia is based on the non-native Pacific oyster. The first Pacific oyster lease in Little Swanport, of 52.7 ha, was granted in 1978. It was valid for 20 years and subsequently renewed in 1992 for another 20 years. A second lease of 10.3 ha was granted in 1983 and a third for 16.8 ha in 1986, both for 20 years, and have subsequently been renewed. The initial development of the oyster industry in Tasmania was ad hoc, with applications for new leases processed individually and outside any formal planning process. There was no sense of considering the cumulative impact of a number of leases or the ramifications for other industries or the environment. As the oyster industry developed, this lack of management in the coastal zone led to potential, if not actual, conflict between the different uses of the environment. The possibility of resource-use conflict came to be recognised by the Tasmanian Government, and it introduced the Marine Farming Planning Act in 1995. As part of this new process, the government devel-

Table 131. The land use characteristics of Little Swanport catchment

Land use description	Land use code (ALUM Version 6)	Area (ha)	Percentage of total catchment area	Possible irrigation/ water allocation
Agriculture-related				
1. Grazing modified pastures	3.2.0	35 166	40.1	Some
2. Grazing natural vegetation	2.1.0	3258	3.7	No
3. Irrigated cropping	4.3.0	105	0.1	Yes
4. Rural residential	5.4.2	79	0.1	Some
Total agriculture (upstream of estuary)		**38 608**	**44.0**	
5. Estuary/coastal waters – production*	6.6.2	0.18	0.0	No
Conservation and minimal use				
6. Strict nature reserves	1.1.1	3143	3.6	No
7. Habitat/species management area	1.1.5	3281	3.7	No
8. Other conserved areas	1.1.7	4190	4.8	No
9. Managed resource protection	1.2.0	42	0.0	No
10. Defence	1.3.1	7707	8.8	No
11. Residual native cover	1.3.3	20 249	23.1	No
Total 'natural' area		**38 612**	**44.0**	
Forestry and Mining				
12. Production forestry	2.2.0	9954	11.3	No
13. Plantation forestry	3.1.0	316	0.4	Some
14. Mining	5.8.0	2.8	0.0	No
Water (non-agriculture)				
15. Reservoir/dam	6.2.0	171	0.2	No
16. River	6.3.0	97	0.1	No
17. Marsh/wetland	6.5.0	31	0.0	No
18. Estuary/coastal waters – intensive use	6.6.0	0.03	0.0	No
TOTAL		**87 792**	**100.0**	

Source: GIS land use maps provided by Mark Brown from DPIW, in 2008 (*pers. comm.*).
* Given their terrestrial focus, the land use maps used as the basis for these data probably do not provide good representations of marine estuarine production activity areas such as oyster farming. More accurate measures are provided in the discussion below.

oped a zoning system for estuarine and coastal waters, similar to land-based schemes. Marine Farming Development Plans (MFDPs) were developed for fish and shellfish growing in waters around the state, which included an environmental impact statement before approval of aquaculture activities, development proposals for marine farming, and management controls and operational constraints for farming activities within the zones.

Today all marine farming operations are licensed under the *Living Marine Resources Management Act 1995*. These licenses include provisions for environmental controls to ensure that marine farming operations are sustainable and do not have an unacceptable impact on the marine environment. Additionally, all bivalve shellfish lease areas must comply with the requirements of the Tasmanian Shellfish Quality Assurance Program, managed under the *Public Health Act 1997* and the *Food Act 1998*. This includes monitoring water quality in shellfish growing areas and the public health status of shellfish on marine farms. These requirements are to ensure the safety of farmed shellfish for human consumption.

The Great Oyster Bay and Mercury Passage Marine Farming Development Plan (MFDP), which includes the Little Swanport estuary, was adopted in October 1998. In Little Swanport estuary a zone of 79.84 ha was allocated with a maximum leasable area of 59.84 ha. This was a reduction of 20 ha from the already existing leased area (the excluded area was largely unsuitable for oyster farming) to reduce the stocking level in the estuary. Sections of oyster leases within the estuary that were outside the new zone were relocated at this time to be within the zone. Part of a then existing oyster lease was also allocated in the Plan as a shellfish nursery area. The Little Swanport MFDP was reviewed in 2008 after 10 years of operation, and minor changes were made to the leases in Little Swanport, including an additional lease of 0.55 ha for nursery culture.

The formal management controls in the MFDP apply to all marine farming zones and are enforceable under the Tasmanian Government legislation that aims to foster sustainable development of the aquaculture industry. Management controls for shellfish stipulate that shellfish farming must not result in any unacceptable impact outside the boundary of the lease area. Furthermore, relevant environmental parameters must be monitored in accordance with the requirements of the relevant farmer's license.

With the rapid expansion of oyster farming in the 1980s some water bodies became overstocked with shellfish, resulting in slower growth of the farmed oysters. After discussions with the shellfish industry representatives, the government introduced environmental controls to manage carrying capacity in shellfish growing waters. These are outlined in Box 13.1.

There are also environmental controls in place in relation to the monitoring of shellfish farms. These specify that all shellfish leases must comply with an environmental monitoring program as stipulated in the relevant marine farm license. Farmers can also be required to provide estimated numbers of each species of shellfish being farmed on the lease. Based on surveys of environmental condition around shellfish farms and a risk assessment of the

Box 13.1: Environmental controls relating to carrying capacity

1. In all new lease areas used for the intertidal farming of oysters there must not be more than 1 km of stocked racking per ha of lease area. When racking is next replaced in all existing lease areas used for the intertidal farming of oysters there must not be more than 1 km of stocked racking per ha of lease area.
2. Containers of oysters in intertidal lease areas must be clear of the seabed and there shall be no layering of containers on the racking.
3. In all new lease areas used for deepwater farming of shellfish there must not be more than 1.1 km of effective backbone longline per ha of lease area. When longlines are next replaced in all existing lease areas used for deepwater farming of shellfish there must not be more than 1.1 km of effective backbone longline per ha of lease area.
4. All longlines and associated equipment for filter feeding shellfish must be maintained clear of the seabed.

Source: www.dpiw.tas.gov.au/inter.nsf/Attachments/PWOD-86A9M7?open, p. 119

effects of shellfish farming on the environment by TAFI in 2000 (Crawford 2003; Crawford *et al.* 2003), the risk of environmental impact from shellfish farming is considered to be low. Thus the farmers have not been required to conduct any ongoing monitoring. Instead, annual inspections are conducted by state government marine farming inspectors. For all new leases and any expansions greater than 10 per cent to existing marine leases, a baseline survey is also required that provides a map of the seabed, photos from each corner of the lease, information on sediment particle size and organic carbon content, and a basic survey for macroinvertebrates.

Thus, marine farmers have considerably more restrictions placed on how they manage their farm than corresponding terrestrial farmers, particularly in relation to controlling environmental impact and stocking rates. These matters are generally left to the sheep and beef grazers to decide on individually, and to their own assessment of the situation. Farming in waters is different, largely because it occurs in water that moves between leases. What one farmer does or does not do is soon experienced by a neighbouring farmer. In terrestrial ecosystems any flow-on of impacts can be greatly delayed. Cultured oysters rely on the movement of water around the entire body of water, and not just the lease area in which they are located, to supply their food. An overstocked oyster lease will impact on all other leases growing in the same body of water. Similarly an environmental impact on one farm is likely to affect other farms in the region.

The production of farmed shellfish has developed into an important commercial activity in Little Swanport estuary. Not only is there direct farm-to-market trade, the shellfish nursery takes newly settled oyster spat from the hatchery and grows them in upwellers until they are large and robust enough to be placed on conventional marine farms. They are then distributed to oyster leases in Tasmania and South Australia. The oyster nursery has produced up to approximately 70 per cent of the spat for on-growing on farms in both Tasmania and South Australia, and the oyster farms have an estimated gross oyster return of approximately $2 million of production in the Little Swanport catchment. Considerable more income is earned along the value-added chain.

Grazing

Over one-quarter of the catchment has been converted to grazing for fine wool merino sheep. A mix of sheep (including fat lambs and wool sheep) and cattle accounts for another 18 per cent of the catchment. In terms of gross income, grazing dominates the catchment economy. The sheep and cattle land is a mix of 'run country', lightly forested 'bush', and completely cleared land, some managed as improved pastures.

Sheep and wool

Given the importance of sheep and wool in the catchment, it is important to elaborate on the role of this industry in both an Australian and Tasmanian context. Australia is the world's major producer of fine (merino) wool, and the state of Tasmania supplies the finest of the fine wool. Australia's climb to the top as a fine wool producer commenced not too long after European settlement of the country in 1788. In 1793 John Macarthur was given 200 ha of grazing land at Camden, near Parramatta. This was to be a turning point in Australian history. Macarthur subsequently bought five merino rams and one ewe from South Africa. Merino sheep had been taken previously from Spain to South Africa. Macarthur was later given another 200 ha in recognition of his successful wool production.

For many decades, until the end of the 1960s, Australia 'rode on the sheep's back'. The industry reached its zenith during the Korean War, between 1950 and 1953. The price for top-quality wool peaked at 'a pound a pound'. During the peak years of wool's dominance, there were 10 times as many sheep in Australia as humans; for example, when there were 17 million humans there were in the order of 170 million sheep. The sheep population is presently relatively low, at around 90 million. Not that long ago the sheep population was approaching 200 million, as the human population approached 20 million.

The rather dramatic changes in sheep numbers reflect the relative returns from other types of farming. Farmers – when they can – respond to changes in product prices (and the costs of inputs); hence, high beef prices and low wool prices will result in fewer sheep and more cattle being grazed. There are many other substitutions farmers need to consider, particularly if the type of land and rainfall gives them flexibility; for example wool country has been converted to grains, and irrigated cotton has made a dent in grazing. However, wool and beef have been the traditional substitutes.

At the time of writing, fine wool prices have improved significantly on what they were recently, and are at a level where reasonable farm incomes can be made with a significant flock. That noted, wool prices are not stable and there have been many shifts, some dramatic, over the industry's history. Since the 1960s, wool has faced competition from artificial fibres. This was a battle many predicted the artificial fibres would win. As wool prices declined in the late 1960s and early 1970s the contest appeared to be over. Yet the artificiality of artificial fibres gradually lost its appeal. It is no longer the conventional wisdom that the fibres made from fossil fuels will win, particularly as oil prices have increased significantly over the past 30 years. The future of wool – particularly if it is grown on rain-fed country and without the need for fossil-fuel-based fertilisers – has not looked so good for decades. What could spoil the party are the increasing costs of inputs (artificial fertilisers and the need to purchase water).

Kirkpatrick and Bridle (2007, p. 45) tell us that a 'typical specialist sheep farm sold 24 484 kg of greasy wool' in 2004, while a 'typical mixed enterprise sheep farm sold 9655 kg'. At 500c/kg (wool prices are given in cents as a convention) this would mean the typical specialist farm would earn a gross income of approximately $125 000 per year. Double the price of wool to 1000c/kg and the gross income becomes $250 000. Increase wool prices further to 1500c/kg and the gross income becomes $375 000. Note, all three prices have prevailed for periods over the past 10 years. The typical mixed-farm would have earned 40 per cent of the above sums from wool sales, plus whatever it earned from other farm products. In these examples, the sheep numbers are not that large. For example, if the average fleece weighs between 5 and 6 kg we are talking about farms with about 2000 to 4000 sheep, small by Australian standards.

Not to present too rosy a picture, let us consider a fairly recent past period when wool prices were poor. Kirkpatrick and Bridle (2007, p. 46) report that the average specialist sheep farm was worth $2 300 000 in 2004, and the average mixed farm $2 500 000. These authors rely on a 2003 economic study that showed a return on capital of 4 per cent in 2000–01 and 5.7 per cent in 2001–02. If we round these out to 5 per cent, we are looking at $115 000 to $125 000 per annum, comparable to the data above for 500c/kg. As stated, the years in question were ones of poor prices. Box 13.2 summarises the situation in the closing days of the first decade of the twenty-first century. The sheep are Tasmanian merinos.

On the basis of the figures in Box 13.2, a flock of 5000 sheep would produce a gross return of $300 000. It should be noted that the price obtained for the finest classes of a clip are in the order of 1700c/kg or more. A price of 2000c/kg is not unusual for top-quality superfine wool. On the other hand, prices for coarser fibre wool, drought-affected wool (weak fibres), and wool discoloured by grease and/or soil are considerably lower. As is the case for belly wool, 'locks' (small second-cuts) and 'pieces' (ragged ends removed from the

Box 13.2: Fine wool facts

- Average price per bale: $1650 to $1800 range
- Average weight of a bale: 140 kg to 180 kg
- Average weight of a fleece: 4.5 kg to 6 kg
- Number of fleeces (sheep) per bale: 27 to 30
- Annual value of wool per sheep (fine wool ewe): approximately $60

fleece) fetch much lower prices. What is important is the average value of a fleece with the low quality wool included. In Box 13.2, a 180 kg bale fetching $1800 means 1000c/kg, which is much lower than the price obtained for superfine wool in recent periods, when prices reached approximately double this amount. Hence, a 5000-sheep property could gross considerably more than $300 000 per annum. However, sheep do not thrive and fleeces do not grow large and fibres strong if food and water are scarce. How many sheep one runs per ha is a function of the grass cover, and that is a function of the water available to the grass. Even the best fine wool country in the world is subject to the vagaries of nature – as we will see. We shall also face the real world difficulty of drawing economic values for both land and water used for wool growing when the price of the final product jumps around over a wide range. We shall come to note that the value of an input into farming (such as water) is determined by the selling price of the final product. This, of course, does not apply only to wool, but also to meat, crops, and aquaculture products such as oysters.

Forestry

In recent times, industries other than grazing and potato growing have come to the catchment. Timber-getting, and in particular woodchipping, is one that has been around for many decades, but plantation forestry, although minor, is a newcomer. The history of woodchipping in the area goes back some 40 years. One of the first coupes in the area to be used for timber is an area of Crown land along the Buckland–Woodsdale Road. It is now regrowth forest. Over the past three decades considerable woodchipping has occurred in the area around Woodsdale. At the back of Mt Hobbs there is a trucking road that joins Buckland to Woodsdale. Along the trucking road the land is under the ownership of the Crown, and the forest to be seen today is regrowth. What is new to the catchment is relatively extensive plantings of native hardwoods in parts of the upper catchment. The plantations have only been established during the period this study has been underway and there will be many years before any income is earned.

The catchment communities

Just as we can categorise the catchment according to the industries it supports, we can characterise it according to communities of people. As will be seen, each community is distinct. The available cadastral maps show there are in the order of 300 separate blocks in the catchment. Most are in private hands, although there are the large areas under the control of the Australian Defence Forces and the Tasmanian Government (as forestry resources). There are a small number of tiny blocks in the hands of various government agencies (for example for telecommunication purposes).

Not all private blocks are owned separately. There are farmers who own two properties, some three. There are small homestead blocks separated from the farm. These factors mean that there are fewer than 300 resident landholders out of a population of 521 people. Many of the coastal residential blocks have holiday homes built on them. Most of the owners of these properties reside in Hobart.

On the basis of the 2006 census, the number of people residing in the Little Swanport catchment was 521. There is no obvious reason to expect anything except the most marginal change to have occurred in more recent years.[1] A significant number of households contain two or more persons; consequently the number of households, particularly occupied ones, at any point in time is much less than the population; it is estimated to be in the order of 200. At school holiday times, or during summer and long weekend holiday times, the number of occupied residences increases considerably.

In the upper catchment, grazing and farming dominate. There is a small number of relatively large grazing properties and a much larger number of small farming units. The large farms are pastoral properties, while the smaller ones can be defined as mixed farming. The smaller farms are centred on Woodsdale. This is the social centre for these small-scale farmers (see later). The existence of a large number of residential blocks in Pontypool and Saltworks Road in the lower catchment means that this part of the lower catchment can be considered to be a totally different community from the rest of it (however, see below). The lower catchment graziers identify with upper catchment graziers rather than with the non-farming lower catchment residents. In this sense there is a graziers' community and a residential community. However, being a significant distance apart – there is no all-weather road from the lower to the upper catchment – means there are two farming communities. There is a further division between the large grazing properties and the small mixed farms.

It is not strictly correct to describe the people living in Pontypool and Saltworks Road – some of whom are permanent, the others weekend or summer holiday-makers – as a community. The thing they have in common is where they live. Their lifestyles and interests differ considerately. They come from a variety of social backgrounds, have different interests, are of differing economic status, and (importantly) have no community, social or recreational centres to bring them together. This part of the catchment is a geographical community – in terms of where people live and where they travel to for goods and services – but it is not a social community, yet considerable informal social interaction does occur.

We can present a precise assessment of the concentration of residents that we have been discussing. The relatively small area in the south-west (south of the village called Whiteford) is the Woodsdale area we have referred to. There are 148 people in this small location. The western part of the catchment has a total of 226 people, which is approaching half the total catchment population. The far eastern part of the catchment has 288 people residing in it, over half the total. Of these, 58 people are in the Pontypool and Saltworks Road communities. Throughout the vast middle section there are only seven people. With this general picture in mind, let us document some key features of the adjacent towns – always mindful that they are not in the catchment. We will start in the north-east.

Swansea

Swansea is a tourist town, catering for holiday-makers who spend a night or two in town, people who have a holiday home in town and are most evident in the Christmas period, and finally passing traffic. It is also a service centre for the local population, including some who live and/or farm in the Little Swanport catchment to the south of the town. The town has a range of accommodation houses, from large motels to B&Bs. It has four 'fine dining' restaurants plus a range of other eateries. It has a large general store, plus real estate agents, a post office, and various other small businesses. There is also a hardware store that is part of the True Hardware chain, owned by the national hardware business chain Mitre 10. The customers are local residents (which can mean people coming from as far away as Coles Bay, a 45-minute drive away) or the Little Swanport lower catchment, a 20–30 minute drive. The hardware products sold in the store are sourced from Melbourne (approximately half), Launceston (about 40 per cent) and Hobart (10 per cent).

The Swansea general store is a combination one-stop shop for groceries, basic hardware goods, fishing equipment, camping equipment, and a few clothing items. There is also a butcher shop under the same roof. The general store has been owned by the same family since 1838. This was not long after Swansea was established and is one example of a very strong 'commitment to place'. A commitment to place (not in a hurry to leave) is a proxy measure of social capital. The supermarket component of the store is today part of the IGA chain. It is a major employer in the town, with nine to ten workers (not all are full-time); all except one were local when we interviewed the owner. Swansea businesses compete with those in Triabunna in the south of the catchment for the custom of residents in the lower catchment.

Swansea is a town with high social capital, which is illustrated by the fact that in 2007 it was awarded the title of Australia's Tidy Town, a prestigious environmental award based on civic pride, partnership, and on-going efforts to maintain a pleasant environment. There is no more coveted civic award in Australia. That such a small town with a small rate-base (and, hence, little spare cash) could win suggests a great deal of community effort. Swansea's small size is illustrated by the fact that it has a reasonably stable population of approximately 600 people. However, the population can nearly treble during summer, with seasonal residents taking up their holiday homes and visitors staying overnight in the commercial accommodation.

Triabunna

Approaching Triabunna from Hobart the road is safe but winding, up hill and down dale. The two high hills one has to conquer have wonderful names, Bust-Me-Gall Hill and Break-Me-Neck Hill. They were given these names in the horse-and-buggy days for appropriate reason. Triabunna is another very old Tasmanian town, established in 1830 as a field station for the 63rd Regiment. You can imagine that period by reference to the still-existing reminders – the police watch-house, built in 1841, and the Springbay Hotel, built in 1861. If the name of the town intrigues you, it is the name for the Tasmanian native hen in the Aboriginal language of the district.

Triabunna is a small fishing port, an export port for wood chips, splintered from the tall timber sourced in the nearby hills, a service centre for the local region, and a holiday destination. Its nearby neighbouring town, Orford, is primarily a holiday centre. Given its function as a small port, Triabunna has a range of businesses that support both the export woodchip industry and commercial fishing. The present population may more than double in the near future if new residential land is opened up. Between 700 and 1000 residential allotments are in the process of being developed as we write. Lifestyle (extremely pleasant views, water-based recreation including fishing, and golf) plus the proximity to Hobart are major drawcards for permanent residents and holiday-makers to the region. A recent influence on Triabunna's economy is the change to the duration of the fishing season. The scallop fishery, which has been the mainstay of fishing in the area, is markedly seasonal. Recently, scallop fishers have switched to harvesting mussels in the off-season. There are between 40 and 50 commercial fishers operating in the locality. A fish-frame rendering plant also has been recently established to produce oils, fish food, and fertilisers. Nature-based tourism is a crucial industry throughout Tasmania and the Triabunna area is part of the picture. The town is the embarkation point for campers and day-trippers heading to Maria Island. The island is an outdoor family camping destination and attracts small numbers of backpackers.

Two features of retail shopping in Triabunna warrant special mention. It has the largest hardware store in the area from Buckland in the south to Coles Bay in the north. Hence, with the increase in residential population, this is an important specialised business. Very little is made locally or produced within the catchment. Only some timber products and cement are sourced locally. Another special feature is the Triabunna butcher shop. It specialises in local meats, sourced from a very small abattoir just outside the upper catchment. There is no other butcher selling local meats. Then there is the Springbay hotel, the hub of the town and, as noted, old by Australian standards.

Oatlands

As explained previously, the site for Oatlands was chosen in 1821 by Governor Lachlan Macquarie. Oatlands is a relatively small historic town in the Midlands of Tasmania, just off the main highway, approximately half-way between Hobart and Launceston. Both these cities are within an hour to two hours drive of Oatlands. This makes this town a tourist destination for overnight stays and stopovers en route. It is also the major service centre for the farmers of the Little Swanport upper catchment, plus many other Midlands farmers from outside the catchment. Proximity to Oatlands means that upper catchment residents and landowners have no reason to shop in the coastal towns. If there is a 'country town' near to the catchment, Oatlands is it.

At an early stage in Tasmania's history, Oatlands was being considered a candidate for the colony's capital. This points to the importance of the early farming industry in the area, and the politically-neutral location between the competing towns of Hobart and Launceston. While other ideas prevailed, the town developed into a picturesque village of some real significance in its early days. It has the largest number of sandstone buildings within a town boundary in Australia. There are 138 of these buildings dating from the nineteenth century. Today, Oatlands has a permanent population approaching 1000 people. It is the seat of local government for the Southern Midlands. It may not be the state's capital, but it stands as testimony to the early history of European Tasmania and the birth of the world's best wool industry.

For a very small town, Oatlands is chock-a-block with small business enterprises and community facilities. The number depends on the extent to which a multi-purpose business (for example a coffee and craft centre, or a hotel and a dining room) is categorised into sub-categories. If separation is precise, there are presently over 130 businesses in Oatlands. In the eatery category (cafes, restaurants, road-houses, hotels) there are 15 enterprises. In health care (excluding the multi-purpose health care facility and the hospital, but covering everything else) there are eight businesses. Seven enterprises are involved in cartage/trucking. And there are seven involved in the sale of old goods and books, with most being antique shops. There are five accommodation houses, generally of the B&B style, and five garden suppliers. There are agricultural contractors, agricultural suppliers/stock agents, earthmovers, bus services, clothing shops, craft shops, and shearing contractors. There are banks, childcare centres, dressmakers, grocery supermarkets and hotels. Finally, there is a wide range of one-operator businesses: a butcher, a bakery, a chemist, a hairdresser, a laundromat, a mechanic, a newsagency, a petrol station, a vet, and a video-hire facility, plus many more not listed here. A variety of community facilities also exist in the town. Pride of place would have to go to the churches, of which there are four; also a town hall and the shire council, a library, a meeting venue, a community radio, and a local newspaper. Oatlands epitomises a small, economically vibrant rural township that came into existence to service farmers and survives today by meeting the dual and contrasting needs of local farmers and the tourists passing by.

14

Communities and values

Tor Hundloe

In this chapter we detail and analyse the economic characteristics of discrete communities within the catchment. We go beyond the purely descriptive of the previous chapter. We account for what is produced and what people do, including what they like (and dislike) about their locality. The data for this chapter comes from our extensive interviews with landholders. This is what scientists call primary data – we were the first to gather it.

We need to say a little here about our data gathering procedure. We attempted a census of landholders; that is, to interview every landholder personally. This was not achieved. In social survey research it is the rarest event to complete a census. Yet such are attempted; in Australia as in many countries a population census is attempted periodically. We did achieve a respectable response to our survey and only the small number of very large farms are under-represented. The response rate for small-to-medium sized farms was 50 per cent, while that for the considerably smaller number of larger farms (wool, fat lambs, cattle) was 35 per cent. The response rate for the non-farming, residential community was approximately 80 per cent. Our difficulty was not so much that landholders did not want to be interviewed, although there were a few in this category, but rather that, try as we might over many weeks of knocking on landholder's doors, we did not manage to meet with a number. Eventually, after failing to meet landholders at the door we left questionnaires and postage-paid envelopes at the residence hoping that those receiving these would become respondents.

The farming community

We start with the farmers. Farm sizes range from just under 200 ha to 6500 ha (the two very small olive farms are excluded). Farms can be categorised in various ways; for example, by size (as above), landforms (improved pasture, run country, cultivated, etc.), or by the products produced. Where a variety of crops and animal products are produced there is a point in describing farms by type of agriculture. The catchment's farm land can be categorised by products produced as in Figure 14.1. It shows that nearly two-thirds of the land is devoted to wool production plus complementary farming of fat lambs and/or beef.

The percentages we report here are not set in stone. They have not remained constant over the years and we should not expect them to do so in the future. The changeability of land uses with varying economic and environmental consequences is a fact of rural life and a serious complicating factor in linking economic and ecological values. However, let us continue. Relative market prices for specific farm produce, and rainfall, determine the

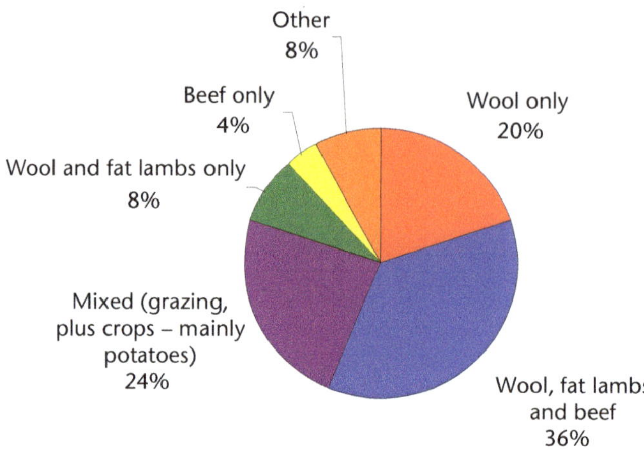

Figure 14.1. Types of farming in the catchment (excluding forestry and aquaculture).

mix of farm production in any year for all but the specialist farmers and those who have a guaranteed source of water. Significant changes in financial returns for a particular crop or for farmed animals will result in farmers changing their crops or animals at the first opportunity. Note that the first opportunity can be years away, depending on the 'sunk' investment in the existing type of farm production – the change from sheep to cattle is relatively straightforward while the change from grazing to crops requires land preparation and the purchase of new farm machinery, which takes time and money. For this change to occur farmers will have to expect the shift in relative returns to be a long-term prospect – otherwise why bother!

During the period in which our study of the catchment took place, the Little Swanport farming community experienced a number of perplexing challenges for farmers. On the one hand there was an extreme drought, and the increased cost of maintaining sheep and cattle in this circumstance. On the other hand, the recent very good prices for fine wool provided the incentive for farmers to stay with fine wool – nevertheless at considerable cost during the drought. Then there was mounting evidence that forestry plantation companies had been offering farmers attractive prices for their properties: the foresters are seeking to meet long-term requirements for plantation hardwoods. In the case where farming has been in the family for generations, the idea of selling the farm is emotionally charged. It is very difficult for farmers to make plans for the future if they are faced with conflicting goals (to remain farming or to sell), and in a period where weather conditions are more unstable than is normal.

Farmers (all farmers) expect droughts to break eventually. The experience of the recent, extremely severe drought in our study area – and fears of an increase in the future in the number and intensity of extreme weather events due to climate change – could (we don't say will) change expectations of the future. Farmers might become more pessimistic. The individual financial circumstances of each farmer are crucial in determining what he or she does today about the future. There is an old adage: if you can survive in the bad times, you'll thrive in the good times. Working to this philosophy, many farmers will strive to survive and see droughts out.

The point is that the vagaries of nature and the vagaries of markets are the determinants of land use patterns. On the basis of data we gathered in our survey, the percentage of farm properties under various types of land cover are shown in Figure 14.2. It is impor-

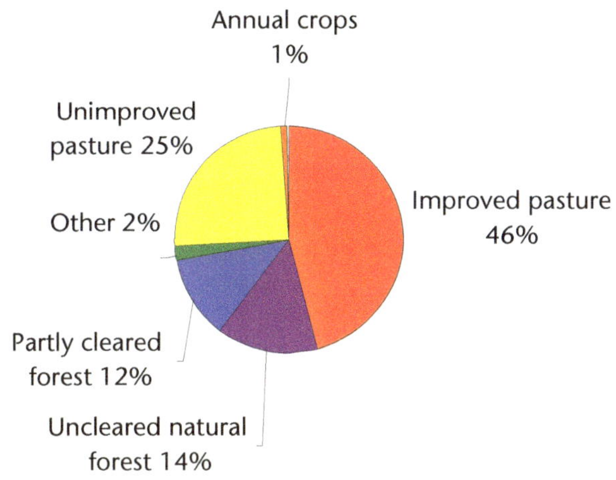

Figure 14.2. Farm land cover.

tant to note that these percentages have changed over time, and they will continue to change as discussed above. This particularly applies to the extent of improved pasture and the area cropped. What we present here is a snapshot. This particular snapshot will help us in our task of calculating the value of incremental amounts of water to farmers, because we are able to take other snapshots of farming in the catchment in a different period, specifically a period of good rain.

The residential community in the lower catchment

There are two distinct, small, compact groupings of residential allotments in the lower catchment, both near the mouth of the estuary. They are known as Pontypool and Saltworks Road. In addition to these settlements there are also a number of non-farm landholdings spread along the main coastal road, which meanders from south to north, never far from the south Pacific Ocean. The road is the Tasman Highway. People living, temporarily or otherwise, on these highway blocks are also part of the lower catchment non-farming community. The non-farm landholdings are typically small. The median-sized holding is 2 ha and the average marginally larger at 2.5 ha. Two-thirds are permanent residents and one-third semi-permanent. The latter category ranges from weekend residents to occasional residents, such as those visiting only during holiday periods or at whim. Of the permanent residents, 50 per cent are retired or semi-retired. Unless one works for enterprises in the catchment (such as hauling timber), the closest work is in the towns we have described.

One of the matters of interest for our research was to discover and document the attributes of the lower catchment that draw people to the area. We gained an understanding of why upper catchment and lower catchment folk farm, but there is no obvious reason for settling in the lower catchment – except its very attractive environment. We set out to understand why these people lived where they lived. In addition to lengthy face-to-face interviews with a small number of selected residents (15 in total), we delivered by hand or by post (depending on whether or not landholders were in residence) a brief questionnaire to every household in the coastal communities. We received a very good response rate of 80 per cent. It is on the basis of responses to our surveys that we can describe these communities and in particular the values and attitudes of the residents. Taking up the research

Table 14.1 Benefits of living in the lower catchment

Nominated by more than 50 per cent	Nominated by less than 50 per cent
Peaceful Good weather State of the environment/clean beaches	Lifestyle Good fishing Access to coast/ocean air Rural setting Good community setting

interests we discussed in Chapter 8, we were particularly interested in two characteristics of the residents' lifestyles: the extent of social capital; and residents' appreciation of, and values attached to, ecosystem goods and services.

Residents described the 'neighbourliness' of the area where they lived as between 'fair to good'. This suggests healthy but not strong social capital. Membership of, or activities with, clubs and associations was virtually non-existent, with involvement in a shooters' association the only activity to attract a number of residents. Maybe there are just too few permanent residents to form viable clubs. Maybe the part of their life of 'being involved' as a daily matter was discarded once they left 'the big city'. Maybe there is not much in common, as a base to be built on, in a 'rural residential' settlement. These are all possible answers.

What we discovered in the lower catchment residential communities tended to be in stark contrast to what we found out about residents in the upper catchment. The latter was a truly engaged community, high on social capital. While there is a lack of community, social and recreational organisations in the two lower catchment communities, there are considerable informal interactions, and hence the degree of social capital is likely to be much higher than the lack of formal connections suggests. What we found here is important for those seeking to understand and document degrees of social capital – it can exist below the vision of the observer. The conventional approach of looking for formal links only goes so far; better to dig further and seek evidence of informal associations that can generate levels of trust and 'looking out for your neighbour'.

Benefits of living in the lower catchment

The lack of formal social opportunities can be easily compensated by other features of the area. The benefits identified by the respondents are listed in Table 14.1. They commence with the most frequently nominated. The first three attributes combined account for approximately half the responses. These identify the combination of a peaceful social environment and a healthy natural environment – not that the other attributes can't be also categorised as elements of these two. One only has to compare this list with the negatives of big city living to understand why people live at the bottom of Little Swanport catchment.

Recreational pursuits of lower catchment residents

Why people reside in our catchment was one issue, and one that we believe is easy to understand. Our next issue of interest was what people did when they were in residence in the lower catchment (which for some was only weekends and holiday time). Did they make the most of the environment's attributes that they lauded? If not, we would have a conundrum. The lower catchment, being adjacent to the Little Swanport estuary and east coast beaches, lends itself to beach-oriented activities. Not only are seascapes available to those who have built with these in mind, fishing and boating are available – at little cost – to all residents. We found that fishing was popular, and to a lesser extent so were other beach and water activities. Away from the beach, tending the garden ranked highly. This is not unexpected. Retired people tend to like gardening. Some get a spiritual feeling in the

Table 14.2. Recreational pursuits of lower catchment residents

More than 50 per cent undertaking	Less than 50 per cent undertaking
Walking (beach, bush, general) Fishing Gardening	Swimming Bird watching Shooting Sailing/boating

growth and blooming of gardens, ever replenishing the soils as plants decay and new life appears. In addition to these activities, the rural environment provides opportunities for activities such as bushwalking and nature appreciation, and these had their devotees. The major activities are shown in Table 14.2, in priority order, with the left-hand side accounting for more than 50 per cent of respondents.

We discovered that there was a large range of activities that were specialist interests of only small numbers of landholders. These include horse-riding, bike-riding, bee-keeping, studying natural history, craft activities, and reading. We also found that those with the inclination engaged in a number of social activities, including visiting friends, playing cards and chess, playing football, pistol shooting, and being active in the Girl Guides. In Table 14.2, walking (a clear first), plus fishing and gardening, account for well over half the recreational activities of lower catchment non-farming residents.

Commercial and semi-commercial activities in the lower catchment

On the basis of our categorisation, any landholder except those whose primary income is earned from farming is a non-farming landholder. This means that there are some businesses – if no more than part-time operations – that fit into this latter category. A very small number of permanent residents in the lower catchment are employed full-time in their own (non-farming, non-tourism) businesses. Unusual enterprises include two very small olive farms. Neither are full-time business enterprises and additional income is earned from other sources. One grower combines olive farming with grazing a small number of sheep and growing grapes.

The categorisation of a landholding as a hobby farm is subjective. Even on small blocks of 2 ha, residents can have substantial vegetable patches and fruit trees; however, this is not the traditional idea of a hobby farm. On the other hand, a slightly larger block of, say, 20 ha where a few sheep are run as fat lambs for domestic use should be considered a hobby farm. Much larger non-farmed areas, of which there are three in the middle of the catchment, are best described as private conservation reserves.

There are two tourism establishments in the lower catchment. One is a property of 150 ha. It provides holiday units, a caravan park and camping facilities, and is a favourite for school groups. Bushwalking, boating, fishing and picnics are popular activities for those who visit. Given its proximity to the estuary, we were interested in the extent to which visitors went fishing, having identified this activity as a potential unpriced use of the environment. The answer is that, on average, a visitor staying one week would go fishing once. A catch of three bream per fisher is the average. The other tourist business is a B&B with a rural emphasis.

Use of the lower catchment natural environment

We were interested not only in what natural characteristics enticed people to become landholders in the lower catchment, but also in what use they made of nature's free gifts (what we have been calling unpriced ecosystem goods and services). It is the usual practice to present people being interviewed with a list of these goods and services compiled by experts – one of the authors has been involved in this practice for many years. The problem with this is it induces respondents to attempt to value characteristics of nature that they otherwise might not have considered important. Lawyers would say we were 'leading' the respondent. Furthermore, there is the potential to exaggerate the value of nature if people think that the interviewer wants something to be valued regardless of the interviewees' opinion. We took the far more defensible approach of allowing respondents to nominate those attributes of nature that they felt benefited them. This we strongly recommend to anyone doing this type of research.

We discovered that some of nature's gifts are essential – as gifts of nature – if one is to live in the area. Water is the main one, as it is not provided as a reticulated service by a government authority, or anyone else for that matter. Catch-your-own is the only solution. Hence, all lower catchment non-farm residents rely on rainwater tanks. There is limited pumping of water from local streams, which is an option for those with riparian rights. Rainwater tanks cost anything from $500 to $5000, depending on their size. However, once the initial expenditure is outlaid, water is free. In economic terms, these catchment residents would value this water at what it would cost if provided as reticulated water by a local authority or state government authority. Water provided by these sources is normally paid for in quarterly payments. Increasingly a two-part tariff is applied – a certain amount is included at a nominal fee to cover water infrastructure; above that a percentage charge applies according to the amount used. Taking into account the cost of the tank, the economic benefit of rainwater to Little Swanport residents (excluding farms) is significant, being valued at a total of $54 000 per year. This has been calculated on the basis of what each non-farm lower catchment resident would be paying for water if supplied to taps in their house and garden as it is in Hobart. Now $54 000 is not much money, but remember there are few permanent residents in the area. If the resident population increased and all had to acquire water in tanks or from creeks, the value of nature's 'free' water would increase accordingly in proportion to the number of residents.

Virtually all the residents rely on septic tanks, and any who don't have septic systems use pit toilets. The latter are few in number and are in the more remote back-blocks. The disposal of waste water is generally onto the garden (if not directly, after a period in a settlement tank), or it goes on to a grassed area or an adjacent paddock.

Another product, free to be gathered for those with enough wooded land, is wood for household fires. Wood fires are commonplace in the catchment (and much of Tasmania). Residents without wood on their property can ask neighbours for it, can obtain the right to take a certain amount from the state forests, or can pay for it to be acquired and brought to the house. For those with enough timber to allow a sustainable harvest for a household fire, the wood is a free ecosystem good, except for the labour and fuel involved in collecting it. A year's supply of 'free' wood is valued in economic terms as the cost of purchasing the equivalent amount. The amount purchased varies significantly across the catchment, depending on the size of the house, the number of the people in it, and whether or not it is occupied throughout the year. The value of 'free' wood to catchment residents is estimated to be $73 000 per year, based on the market price of the same quality of firewood. Again a small amount, but appropriate for a small population.

Fishing

As it can be seen from Table 13.2, fishing is an important activity for locals. Approximately 70 per cent of lower catchment residents fish in the estuary. The same percentage, but not necessarily the same people, fish offshore, where different species are sought and caught. The average number of days per year fishing in the estuary was found from our survey to be 22. The number is large due to a small percentage of frequent fishers. This is the usual situation. A few dedicated fishers pull the average up and disguise the fact that the majority are very occasional fishers. Three-quarters of fishers spend 12 days or fewer per year fishing in the estuary, with the average being six days per year for those people. The overall range of fishing days is 2–90. This is a good illustration of the extremes found in fishing studies worldwide. Fish caught in the estuary are generally confined to two species, flathead and bream. The average catch is four flathead per trip and three bream per trip.[1]

Recreational fishing is both an ecosystem good (the fish caught are eaten) and an ecosystem service (the satisfaction of fishing – 'utility' in the jargon of economics). The appropriate economic measure of a free activity such as fishing is (notwithstanding the cost of any arbitrarily set license fee) the willingness-to-pay for that activity. Economists have to rely on so-called 'surrogate' or 'hypothetical' market information to derive a dollar value.

The former relies on a method called 'travel cost'.[2] There are far too few fishers using the Little Swanport estuary to make use of this valuation technique. The alternative methodology (a hypothetical market approach) relies on a large number of interviews of fishers using the 'contingent valuation method' or 'choice modelling'.[3] These approaches have been used extensively in recent years in Australia (and elsewhere). Likewise, the numbers involved prohibited using the method.

Two points should be noted about the economic value of recreational fishing. The first is that willingness-to-pay for the first fish caught is much higher than for the second fish caught, and so on as more are caught - what economists call 'diminishing marginal utility' sets in immediately and is stark. The second point is that there have been a number of research studies that show that the value of another fish to the recreationalist tends to equal the dollar value of the same species caught by a commercial fisher. On this basis we have estimated that the lower catchment resident who goes fishing obtains an economic benefit of $700 per year. Considerable time might be involved and bait and gear purchased, but by and large fish are a free ecosystem good. In total we estimate the 'value of fishing' to be $140 000 or thereabouts per year. This, as with other ecosystem values discussed above, is a small sum; however we are dealing with a small number of fishers. It is important to note that this has no relationship to what these fishers paid for their bait and gear, but rather what they would have paid for 'the right' to fish if that had been a government requirement.

Given the proximity of the lower catchment to Hobart – and the fact that it is on a very picturesque coastal road – it is not surprising that visitors stop in the area, particularly in the vicinity of the estuary. While not part of the community, in that they do not own land or have business interests in the catchment, visitors use the resources of the estuary. The Little Swanport estuary is genuinely beautiful. The waters are clear and the upper catchment hills provide a frame to the variety of blue shadings of the water and the light green of the low-lying fields of remnant forest. To the east (across Great Oyster Bay) is the world-renowned Freycinet Peninsula, and to the south Maria Island, a centre of interest for both its history and its natural beauty. Visitors to the Tasmanian east coast traverse the Tasman Highway, crossing the Little Swanport River at the western extremity of the estuary (or the eastern extremity of the river). Magnificent views are there for free, all the way from Swansea in the north to Orford in the south.

Unless visitors stop and engage in recreational or tourism activities within the catchment, they are not of direct interest to a study of the catchment. Given that very little commercial accommodation is available in the catchment and only one camping ground exists (although considerable informal camping occurs), there is insignificant visitor impact of either an environmental or economic nature. Those who stop to engage in recreation are most likely to be on day trips or passing through and stop to 'boil the billy' or otherwise 'refuel and refresh' themselves. On the other hand there are those visitors who do stop overnight in Swansea and/or Triabunna/Orford and consequently make a significant difference to the economic health of these townships. Their expenditure is of indirect benefit to the catchment residents, as they help establish the viability of the businesses in these towns and thus ensure the towns are able to provide a range of services that otherwise might not have been the case. Considered in this light, if these townships did not exist, or were much smaller, the lower catchment residents would need to travel to Sorell or Hobart for many goods and services. This would be both inconvenient and costly.

The visitors of direct interest for this study are those who specifically come to the estuary for recreational purposes. They do have the potential to use (or abuse) the environment, and they are likely to have some economic impact. Day trips from Hobart are popular in summer, particularly when the weather is suitable for picnicking on the banks of the estuary and for fishing. Some will swim in the safe waters. Most fishing is done from boats brought by trailer to the launching ramp constructed on the northern shore of the estuary. However, most fishing by visitors is undertaken in Great Oyster Bay, outside the estuary.

It is important to note that recreational fishing by visitors is an important activity in, or immediately outside, the estuary at certain times of the year (summer, Easter and school holidays). For much of the year no recreational fishing by visitors occurs. On the busiest days there are likely to be no more than two dozen boats. Recreational fishing in the estuary itself depends on tides and weather conditions outside. As to fishing further upstream, all we were able to gather over three years of study was anecdotal evidence. There is no reason to doubt that fishing does occur upstream of the estuary and, in fact, good reason to believe that certain 'holes' would tempt anglers. Commercial fishing does not occur in the estuary, although commercial fishing is a major off-shore industry. The boat ramp was upgraded near the end of our field work and recreational fishing has increased as a result.

Ranking of environmental health

So far we have used our survey data to describe the activities of lower catchment landholders and what they like about the area. We did probe deeper, in particular seeking data on environmental and social attributes within the catchment. We were looking for evidence of social capital and use of ecosystem goods and services that might have been missed.

Residents of the lower catchment were asked to rank a number of environmental and societal values of the area. The rankings were over a five-category scale – 'excellent' to 'terrible' – plus the option of a 'not sure'. The rankings are of course perceptions, not scientific measures. However, perceptions are extremely important as they influence behaviour. For example, if I believe I won't catch a fish because the water is polluted or others have beaten me, I'll probably not go fishing. If my perceptions are wrong, I have forgone a pleasurable and probably rewarding activity. Perceptions with regard to environmental quality should not be compared to scientific assessments or professional opinions about the state of the environment. It is common to find that lay opinions (that is, perceptions) can be far removed from scientific reality. Treat perceptions as simply that, but note that they influence behaviour.

Perceptions

Having made this point, it is worth keeping in mind that for much day-to-day decision-making what matters is a person's perception. An unskilled angler might reach the conclusion that the fishery is depleted when it is not. The simple fact is that the angler simply cannot catch fish. However, the angler does not recognise that as the problem; rather he or she blames, say, commercial fishers or polluting industries for his/her lack of success. Take another example. Those without botanical skills are likely not to notice that a forest area has been subject to selective logging in the past, believing that the present assemblage of trees is natural. Perceptions, unless corrected, determine political attitudes and actions. Many environmental disputes are driven by perceptions rather than facts and this is a great pity – and an enormous frustration for environmental scientists.

Ranking of the community spirit and the environment

With regard to ranking of community spirit, nearly 60 per cent thought it was good, one-quarter thought it was excellent, and 15 per cent thought it was fair. On this basis we could conclude that social capital was intact. In terms of the general condition of the natural environment in proximity to where respondents lived, 50 per cent said it was good, one-quarter said it was fair, and one-quarter believed it to be excellent. With regard to the state of the natural forests, over 45 per cent believed they were in good condition, over one-quarter said in fair condition, and about 15 per cent excellent, with another 15 per cent saying poor. This was the most diverse range of opinions and reflects differing understandings as much as anything else as to what good or poor quality forest is. On the basis of these responses, we could only conclude that respondents were generally pleased with the quality of the environment, particularly in near proximity to their landholding. The environmental health of the estuary was dealt with in a similar fashion. Nearly one half said it was good. About one-quarter said it was fair, about 15 per cent said excellent, about 8 per cent said poor, and 2 per cent were unsure. Again there was a diversity of views. In summary, on these four attributes (community spirit, the natural environment, forests and the estuary) the prevalent view was that all were in good shape. Where there was strong conflicting opinions, the contrary views were held by fairly small minorities.

These findings influenced our research approach on the matters of valuing non-market attributes, particularly environmental ones. We had intended to utilise one of the so-called 'non-market' techniques to estimate values of environmental attributes. The strong agreement that the important environmental attributes were in good shape meant that it would be difficult to utilise 'stated preference' techniques to elucidate economic values. A stated preference is a willingness-to-pay measure for more or less of some valued state of the natural (or social) environment, usually measured by contingent valuation or choice modelling techniques. To use these methods it has to be possible to construct believable scenarios based on strongly contrasting views – and if the latter do not exist, it is near impossible to obtain realistic monetary values. We were forced to discard all approaches based on standard measurement of non-market values. However that was not the end of the matter.

Better or worse?

Landowners were asked if they believe the natural environment had changed (improved, worsened or remained the same) in recent years. While little can be made of responses in a scientific sense, any reported changes would alert us to investigate what might have occurred. Over 50 per cent believed there had been no change, while approximately one-quarter considered it had gotten worse, about 20 per cent that it had improved, and the remaining 5 per cent were not sure. Yet again a range of conflicting views. We hope the reader is starting to appreciate the difficulty of coming to a common judgment on environmental matters where we are dealing with attitudes, perceptions and lay interpretations. This is not a trivial research finding. Far too many environmental disputes are driven by ignorance.

Landholders were then asked about changes in wildlife numbers in their local area, compared to five and ten years ago. The responses were that there was no significant difference between observations of animals in the two past periods (five or ten years previously). However there was a difference between the past and the present. Birdlife was considered to be doing best, with 65 per cent saying that numbers had either stayed the same (40 per cent) or increased (25 per cent). Fish presented a different picture, with 40 per cent saying that numbers had stayed the same, and 35 per cent saying numbers had decreased. Ten per cent believed fish numbers had increased and 15 per cent were not sure; all somewhat confusing without empirical evidence to rely on. On the basis of respondents' opinions we could not, even trying very hard, construct tests that could indicate the environmental values at stake.

Property values and lifestyle benefits

Let us turn to another approach in our attempt to understand (and value) the environment. What people are willing to pay for blocks of land, particularly small ones that are used for rural-residential purposes, can be deconstructed to obtain the environmental component of the price paid. In nutshell, aesthetic values as reflected in ocean views (or mountain views) translate into monetary terms. Recent property sales in the lower catchment show people are willing to pay good money for small blocks – particularly those near to the estuary.

When the attributes of natural environment are part of the attraction of a location, the economist will hypothesise that a premium is paid for the environmental characteristic – nothing other than commonsense! If a sufficient number of sales are made over a short enough time period, the economist is able to model the results and search for the actual quantum of the premium. This technique is called 'hedonic pricing'. Unfortunately in our case, even going back over sales data for a five-year period, there have not been anywhere near enough sales to permit modelling of this kind. Go back in time too far in a search for a large number of sales (and in our catchment we would want sales over many decades because of the small number of properties sold per year) and there is a strong likelihood that variables other than aesthetics and the state of the natural environment will play a major part in changes to real estate prices. The general increase in wealth, population growth and 'the sea change' phenomenon are likely to swamp environmental attributes. Looking for an environmental premium on prime locations took us nowhere in the Little Swanport catchment study.

In the catchment, change in property ownership has been insignificant over a long period. It is not unusual for landholders to have owned their block (and house) for over 20 plus years. While not possible to use in our present research, hedonic pricing evaluations are likely to be useful in the future, as Swansea, Triabunna and Orford grow in population.

Table 14.3 lists the natural products that catchment landholders utilise, plus our estimate of their value (based on survey responses). These are things we can value. We have not discussed valuation of 'lifestyle benefits'. However, as the survey results indicate, they are a major feature of living – and working – in the catchment. The sum of $1.5 million in Table 14.3 is an estimate of lifestyle benefits measured as forgone income, which is the difference between the net income a typical catchment farmer gets and the considerably larger amount he or she would get in paid employment if living in Hobart or Launceston. This opportunity cost (or forgone benefits) of farming in the catchment is a monetary reflection of the lifestyle benefits the farmers obtain. It is a significant sum for such a small population. On a per capita basis, nature's small gifts are not insubstantial. Retired people living in the catchment are excluded as they are not forgoing income in a job in Hobart or elsewhere.

Table 14.3. Use of nature's products

Product	Extent of use (%)*	Value/year	
Wood	100	$73 000	Market price
Household water	100	$54 000	Market price
Fish	70	$82 400	Willingness-to-pay
Game	12	–	–
Mushrooms	minimal	–	–
Wild flowers	minimal	–	–
Environmental lifestyle benefits	100	$1 500 000	Opportunity cost

*100% means all respondents used wood from the catchment

The upper catchment

We now turn our attention to landholders and residents in the upper catchment. The major land use of the upper catchment is farming, mixed farming in the Woodsdale area and fine wool, or wool, lambs and beef elsewhere. There are a number of non-farming residents on small blocks and they are spread throughout the area, although they are generally situated along the main roads. The non-farming upper catchment residents are different in a number of respects from those of the lower catchment, as we will see.

Non-farming businesses

In the upper catchment there is a small group of residents who are employed outside the catchment (for example, teachers and tradespeople). Another small group of people living in the upper catchment operate their own businesses – other than farming – in the catchment. Some run farms in conjunction with their other business or professional life. The businesses include a B&B/backpackers establishment, a builder contractor, a graphic design business, a saddlery school, a stock agent, a timber mill and a wood merchant. There are at least three (all female) wool classers.

Social and sporting clubs in the upper catchment

Given the long history of land occupation in parts of the upper catchment and the fact that the descendants of the original farmers are still in the area, social links could be expected to be strong. And so they are. A small number of social/recreational clubs exist, and prosper, within the upper catchment. These clubs are part of the glue that binds the upper catchment community. The clubs are St Matthia's Anglican Ladies Guild, Woodsdale Football Club, Woodsdale Hall Committee, Woodsdale Indoor Bowls Club, and Woodsdale Recreation Ground and Club Rooms Committee. All have a unique function, although there is overlapping membership. Football (of the purely Australian variety) is played by anyone keen enough – the standard is good for such a small population base – and supported by all.

As with the lower catchment community, we wanted to understand what residents liked about living in the upper catchment, what they did for recreation, and where their economic links were. Much of this information we gathered in our questionnaires, but the small number of lengthy, in-depth conversations with residents put the flesh on the raw data.

Living in the upper catchment

The benefits identified by upper catchment respondents are listed in Box 14.1. Combined, these suggest 'country living', which, of course, is what they are doing. The typical recreational pursuits are listed in Box 14.2. All are equally important. The list excludes social and

Box 14.1: Benefits of living in the upper catchment

Fresh air
Quiet
Privacy
Ability to keep a variety of animals

sporting activities, which are the bedrock of the upper catchment community and warrant separate discussion.

Social activities and social capital in the upper catchment

The Woodsdale area is what we can term, and what farmers themselves refer to as, a 'traditional farming' community. This community has strong social capital. People live close enough together to be day-to-day neighbours, and this helps to develop and maintain relationships. This is how it was when the first permanent settlers came to the area, and as it is today. As we discovered, descendants of most of the pioneering families are still in the area. The necessity of neighbourly cooperation in the early days has continued to the present. This has been helped by considerable intermarriage.

The properties are relatively small, with anything over 400 ha being large. Many are much smaller than 400 ha. As mentioned previously, there was a significant dairy industry in the Woodsdale area for many decades and the landholdings were small in that era. Dairy herds were measured in dozens of cows, not hundreds as modern-day herds are. Today, the area is defined by its mixed farming; for example, a farm might have up to 1200 sheep, and 20 to 30 cattle, with a 50:50 split between wool and fat lambs. There are as many ways of splitting the mix of grazing animals as there are farms. While running sheep and cattle dominate, there are other types of farming, in particular potato growing, for both the fresh food market and seed production. There is a small amount of long-established plantation forestry and very recent eucalypt plantings on a farm leased by Gunns. Various other farming activities have been tried in the past (poppies, beans, pigs) but have been discontinued.

Locally based sporting teams and community organisations play a significant role in the life of the residents. There is a mixing of the two. This is what makes the Woodsdale area a genuine community with high social capital. There is a country cricket club, the Levensdale Cricket Club. The cricket club has been in existence for over 100 years. This is worth re-reading. We are discussing a very small, rural sporting club that has a history comparable to some of the giants of Australian sporting history (such as the old Melbourne-based AFL clubs). This is an example of the bricks and mortar of social capital which exists today only in small rural communities. After the population explosion in

Box 14.2: Recreational pursuits of upper catchment residents

Walking
Bird watching
Picnicking
Cycling
Swimming

Australia after the Second World War, the suburban football clubs – still carrying suburban names in Melbourne – saw their membership fracture in geographical terms and spread city-wide. Cricket is played in the upper catchment over the 'summer' half of the year. Games are played at 'home' and 'away' grounds, with every second week being a 'home' game. Training occurs on every Thursday during the season, and every second week there is a barbecue for players and club members. Other less frequent social events are held, such as a 'Santa Day' in December. Such is the spread and depth of local sport.

The Woodsdale Football Club has been in existence since 1954, a relative youngster in local sporting history. In the past there were seven or eight teams in the regional competition, a remarkable feat for a very small catchment. In recent years, with the introduction of weekend trading by businesses in Oatlands, both players and supporters have been drawn away from football – if you are rostered to work on Saturday, you can't be chasing leather! In the past the rules for the regional competition required each club to field two teams, a 'firsts' and 'seconds'. For two years Woodsdale could not meet this requirement because of the loss of players who had to be at their place of work, and hence the club was excluded from the competition. This rule has been revoked and Woodsdale has returned to the competition, where it continues to perform well.

The Woodsdale football team trains once a week and plays each weekend during the season, which approximately equates to the half of the year not taken up by cricket. There were 55 members (players and supporters) of the club when we interviewed club officials, however with non-registered followers of the team included about 100 people are involved in some way or other with the club; yet again a small number, but not compared with the few hundred who live in this part of the catchment. The local council provides the playing field while the club owns the club house and kiosk.

The other recreational groups centred on Woodsdale are a bowls group and eight-ball group. There are two important social organisations, the church group and the Hall Committee. The local church had a congregation of 12 people in 2007. It is serviced by a visiting minister. The Hall Committee maintains the well preserved community hall. It finances this by selling lunches to tour groups who come by bus from either Hobart or Launceston and who make a stop at Woodsdale. There are eight to ten tours per year, with full or near-full buses, the average number of tourists per bus being between 40 and 50 people. Farmers donate meat for the meals and the women make the lunches. Hence, this is a profitable business for the Hall Committee, and the hall, a reminder of a past era – country dances, arts and crafts exhibitions and concerts – is perfectly preserved. Some would say that it is a shame that the need for the hall has been dramatically reduced due to the intrusion of television into the social life of otherwise outgoing people. The global village is destroying the local village, if it has not already done so.

In summary

On the basis of this description of the various separate communities in the catchment we have a base on which to develop our understanding of it as if it were an unified economic unit – a province or even a tiny nation – producing goods and services for domestic consumption and for trade, while importing both the household goods and farm-production inputs it requires to exist as an entity. As already explained, it is a community without any towns within its borders, and hence trade with the 'outside' world is an essential feature of catchment life. To understand our catchment better in economic terms we are going to start with an estimate of what economists call 'the regional product'. This is equivalent to a nation's national product, only at a much smaller scale.

15

The catchment regional economy

Tor Hundloe and Peter Daniels

A catchment's surrounding environment is not limited to towns and cities on the periphery of the catchment. It is likely that a far wider, much larger economy is associated with the catchment community and its economy. That wider economy could be the state, province or nation in which the catchment sits. Recall our discussion of the Nile River basin – the basin is Egypt, the rest is desert. Egypt would not be a country without the Nile. The Mississippi–Missouri River basin in the United States is not all of the United States, but nevertheless a great slice of it. The United States would be a different country – smaller, weaker and probably unrecognisable – without this river basin. How things have changed from the time of Huck Finn and Tom Sawyer! In the twenty-first century, the global economy subsumes (consumes!) all but the tiniest catchments. Little Swanport catchment is a tiny one, but read on to discover that it has important relationships to the outside world.

In the case of our catchment, the global environment, which we should note immediately includes Chinese wool buyers and the fashion houses of Milan, is crucial to the catchment's economic health. The world's best wool (either produced in our catchment or next door in the Tasmanian Midlands) is bought by buyers from Italy, to be woven into high-priced suits, skirts and cardigans. Less valuable wool, but still of very high quality by world standards, is bought by the Chinese, who are today Australia's main wool consumers. Take wool growing out of the catchment and its economy would be unrecognisable – very much poorer, unless it was immediately replaced by an equally profitable substitute, which could be beef and fat lambs if meat prices remained strong.

Good quality meats (both lamb and beef) and wool find their way into numerous overseas markets where what economists call 'world prices' prevail. Whether wool or meat, if you take into account the costs of transport between Tasmania and an overseas market, the local product will find its way to the highest yielding (most profitable) market. This means that world prices are the yardstick for some of the catchment's most important products. This puts the farmers at the mercy of the international economy and movements in exchange rates. Just as importantly, our case study farmers and other businesses have to pay world prices for inputs such as petrol, diesel, and artificial fertilisers. On the other hand, there are local products such as potatoes (heavy vegetables to transport) that find sufficient demand in Australian markets and that would not be considered for export.

In what follows we will confine our analysis to the economic interdependencies between the catchment landholders and the local economy. This will present a realistic and readily understood measure of the economics of the catchment, but keep in mind that both wool and meat grown in the catchment are destined for distant markets. In the case of wool, all the value-adding occurs overseas (for example, milling in China, and design in the fashion houses of Europe).

Be mindful of our fundamental task, which is to analyse the use of water in our catchment. Economic production statistics are vital for identifying the source and the extent of demand, pressure and competition for water in the catchment. They are also a major part of the water productivity indices that can be used to examine and compare various forms of output per unit of water used. The economic structure of an area can be based on the physical output (e.g. kilos of wool, meat, potatoes, etc.) or the related monetary value of that output (equal to physical quantity multiplied by price per unit). The latter usually receives more attention and is often easier to obtain.

While the average citizen might think that there is an enormous amount of economic and personal data 'out there' to be tapped into by governments – something some citizens worry about – this is not the case. Researchers, as we are, find that there is a paucity of data on which we can make useful assessments – and we stress useful – for the community. Decisions, whether made by the government or the private sector, can only be as good as the data that is available. There is limited information that can be used to compile agricultural or other economic output statistics for the catchment. It is worth noting this matter, for no other purpose than to indicate that even in a data-rich country like Australia valuable research is constrained if governments or public-sector researchers do not obtain the cooperation of citizens in providing data.

The best existing source of data for our type of research is the ABS's Agricultural Commodities: Small Area Data (Cat No. 7125.0), but this only resolves down to Statistical Local Areas that are very large, and it cannot be made to match the Little Swanport catchment system boundary. Some useful small-area (enterprise-level) data do exist from the 2001 and 2006 agricultural censuses administered by the ABS. However, they are not stored for retrieval at the catchment level, and special, rather expensive, customised data runs would be required to estimate selected area production and value statistics for our catchment. These were only available for 2005–06 when we undertook our analysis.

Sheep

Sheep, and the production of fine wool in particular, dominate the economic land use of the catchment. While the exact number of sheep and value of wool production varies on a year-by-year basis, it is possible to arrive at a general level. One simple, fairly reliable approach is to multiply the area of grazing pasture by the dry sheep equivalent (DSE) per ha. This is the measure for maintaining a 48 kg wether for a year. It varies according to the productivity of the land. In our catchment the estimates of grazing land vary quite widely – depending on the source, from 19 per cent of the total catchment area (Resource Planning and Development Committee 2003) to 44 per cent based on the land use GIS map (see Figure 13.1). Such a wide variation results from different categorisation of land and the condition of the land at any point in time, the latter being a function of rainfall and stocking rates. We rely on the GIS map and hence settle on 38 000 ha.

Now to the estimates of the DSE for the catchment. A number of expert bodies are involved in calculating DSEs for different sheep-grazing areas in Australia, and at different points in time. Estimates vary from 1.5 to 3 DSE/ha (in the National Land and Water Audit in 2004) (National Heritage Trust 2004), to 4 DSE (Tasmanian estimates from DPIWE 2003), to 5 DSE/ha for the best sheep farming country in Tasmania. The regional DSE of 4 DSE rather than the higher figure for the prime wool country can be considered the relevant measure for the Little Swanport catchment. If we adopt this, then 38 000 ha × 4 DSE = 152 000 sheep. Other indicators suggest this figure may be higher. For example, calculations based on stock water of 350 ML per year, based on 6 L/day for 365 days, suggests 160 000 sheep. However, 150 000 is considered a reasonable maximum estimate and is utilised in our water accounting analysis presented later. Given that approximately 33 sheep are required for a bale of wool per year, and a bale of wool on average fetched $1700 in 2007 (the base year for many of our calculations), this infers that wool production in the catchment would be around $7.7 million per year (at this level of production and with this average price).

Other farming

Potato production is a very minor activity on a catchment-wide basis but obviously important for the small number of farmers engaged in it. Production varies from a low figure of 15 ha, and water demand is minimal. Forestry production is difficult to estimate. While it had been considerable in the early days, it has not been significant recently. That is likely to change when the new plantations reach maturity.

There are few specific data on other agricultural output from the area. Existing economic statistics simply do not have appropriate product, spatial and temporal coverage from which to identify catchment-specific production. Special runs on enterprise-level ABS Agricultural Census data would be a possible source of information. Based on the catchment rural landholder survey, more detailed description of farming by property type is provided in a following chapter.

Measuring what is produced

Economists have, for more than 50 years now, used a method of measuring what is produced, where it goes once it is produced, what inputs are used in production and where they come from so they can model national and, from the 1970s, regional economies. The formal name for this type of model is an input-output table or transactions table. The data are dollar values of goods and services that are bought and sold in markets. No non-market goods and services are in the table. This means much of what is of environmental concern is not captured. Not only does a transactions table model transactions within or between economic sectors of the economy, but it also includes the exports and imports of the economy under investigation. We will use this input-output idea to deal with aspects of the Little Swanport economy, such as its size and its relationships to its nearby trading centres. It is these trading centres and their businesses that flourish, or decline, in response to the health of the Little Swanport economy.

It is not the first (or direct) sale or purchase in a local town that defines the total economic impact of catchment trade. For example, the money spent by catchment residents in neighbouring towns becomes the income of the townsfolk, and the townsfolk go on to spend (much of) it rather than put it under the mattress. Their spending means additional sales and additional income. A multiplier effect is an operation once there is an initial injection of money (such as this season's wool cheque). An ever-diminishing spiral of spending takes places. This flow-on benefit can be represented by output, income and employment multipliers, and is calculated by mathematical modelling using matrix algebra applied to the data in the transaction table.

A matter of interest in understanding this type of economic impact analysis is where we draw the boundary around an economy. For the study of a national economy the boundary is self-selecting, and this is also the case if a state or provincial economy is under scrutiny. If a city or formally defined local government area is the subject of analysis, again we would have defined boundaries. The matter of defining economic boundaries only arises where the economy under consideration does not recognise political boundaries. As we noted in the opening pages of this book, this is most likely the case for catchments. We have already made clear that we wish to analyse the catchment's economy, but we want to do that in a manner that illustrates the catchment's economic importance to its major linked centres outside. It is for this reason that we spent considerable time in documenting the businesses in the nearby towns. We certainly want to include them in our wider economic boundary.

The relationship of the economy of the Little Swanport catchment to the outside world includes the service centres (the towns) on the catchment's periphery, the Tasmanian capital city of Hobart, the rest of Tasmania, the rest of Australia, and finally the rest of the world. Even the most parochial Taswegians would hesitate to assert that Tasmania, let alone a fraction of it, mattered greatly in a global context. And that is the case, yet the global reach of this tiny catchment would register if we cared to trace through to the world's fashion houses the wool that was grown in the catchment. As a matter of interest, if we were studying the central north of Tasmania where most of the world's legal opium poppies are grown, we would discover another small part of Tasmania with an important influence on the planet.

In describing the catchment's economy the emphasis will be on the catchment's farmers, oyster producers, and permanent and semi-permanent residents. They are the source of the economic activity in the catchment. They not only provide a living (and lifestyle) for themselves, but they also play a very significant role in the prosperity of the nearby towns, as well as, in a minor way, the economic health of Hobart and Launceston. By the time wool from the catchment reaches China the economic impact (of the relatively few bales) is lost in the vastness of the Chinese economy. The multiplier effect diminishes quickly as soon as we move more than one step away from the catchment. This is why our focus is local.

Keep in mind the minute size of our catchment. It certainly produces world-class wool, oysters and meat, but none in large quantities. The catchment's very high quality products would still be available in significant quantities if all the producers deserted the catchment, since there are producers of these goods elsewhere. By far the most valuable export product, fine to superfine merino wool, is also produced in the adjacent Tasmanian Northern Midlands, the New England district of New South Wales, and a few other places. Top quality and high-priced oysters destined for the domestic market are produced in nearby Tasmanian waters as well as in South Australia. A different species of oyster (an Australian native, the Sydney-rock oyster – originally the Brisbane rock oyster) is produced in warmer climates in New South Wales and Queensland.

The economic modelling approach that is used to trace through the multiplier impact requires an accurate recording (in monetary terms) of what is produced (for example, fat lambs, potatoes, wool), and whether this product becomes an input in a value-adding industry (such as wool being sent to a woollen mill) or goes direct to consumers (as might farm produce sold from a roadside stall). The approach also requires an accurate recording of the monetary cost of all inputs to the various industries in the economy. In our Little Swanport economy the inputs are what we can call 'farm supplies' (fertilisers, pesticides, fuel). From what other 'economy' do these come? In our Little Swanport catchment, the closest 'other economy' comprises the nearby towns (for example Oatlands). Some inputs will come from far afield, from overseas as we discover if we trace them to their ultimate source; however there is likely to be value-adding as these products move along the chain from foreign producer to Australian importer, national wholesaler, and to local retailer. Such is the hierarchical nature of modern economies. Many get to share – even if only marginally – in the economic game.

The imports and exports of our catchment are not limited to the agricultural sector, as the people resident in the catchment purchase most of their foodstuffs, and all their clothing, entertainment services and the like from outside their 'economy'. We can note that many of the small-scale farmers supply their own meat and some vegetables and fruits and this has a significant impact in reducing individual grocery bills. This will be evident when we present data on household expenditure. We will note smaller than expected spending on food by these farmers. The same situation is found in many agricultural areas where the farms are relatively small. Because farm-grown foods are not sold in the market, their value is not recorded in conventional economic data, except for where farmers are required to document home consumption of agricultural products.

Goods and services purchased by consumers are termed 'final demand' in an input-output model – final because the consumer is the last point in the chain, which starts with growing the wheat, through milling it, to making the bread that will be consumed as toast at breakfast. The 'household sector' not only consumes but also provides much-needed inputs to the catchment economy, in particular labour and human capital to the 'producing' sectors of the economy (such as agriculture, manufacturing, and services such as teaching). The economic measure of a teacher's work is interesting enough to explain. A resident of the Little Swanport catchment who travels outside the catchment for five days a week to teach in a school in Oatlands is 'exporting' a valuable service and the catchment (where the teacher lives) receives payment, the teacher's salary, for its export. The teacher's salary becomes part of the catchment's income.

Tasmanian input-output tables

There was a period – from the late 1960s to the mid 1980s – in which Tasmanians were well served with reasonably rigorous economic data presented in statewide transactions tables. The whole state economy was modelled using this method. Taking the state tables as a starting point, it was possible to construct regional input-output tables for parts of the state by gathering region-specific data and augmenting the state transaction table. Using the existing tables as a starting point, the Little Swanport catchment could have been studied as a region. In this study we don't have that option, as we explain next. Those who make decisions about what type of data are gathered and modelled do not – and to be fair, cannot – know what demands future researchers will make. We were doing our research in the wrong era.

The most recent input-output analysis was published by the Tasmanian Department of Treasury and Finance in 1990. That was more than 20 years ago. The analysis was formulated using 1985–86 data. How far out of date would we be in using that data? While there are often 10-year gaps in the preparation of input-output tables and hence usually somewhat dated data are used to interpret today's economic world, a 20-year-old analysis is not of great relevance today. Too many changes have occurred – both product prices and input prices have changed (some such as fuel, dramatically) and the structure of the economy has changed. The oyster industry as we know it today did not exist. The lack of a baseline, state-wide input-output transactions table implied that we were not going to be successful in modelling the catchment economy and its regional flow-on benefits.

While we certainly couldn't use the conventional method of augmenting an existing state-wide transaction table, our alternative was to build the Little Swanport economy from the ground up. With a very small economy, such as ours, this is not an impossible task. However, it is another matter to calculate the so-called technical coefficients on which the output, income and employment multipliers are based. The practicality of filling a significant number of cells in an input-output matrix escaped us. This was due to the very small transactions involved and limited structure of the regional economy. We were attempting to construct a model with too little data. Just when it seemed we were to fail in this part of our investigations we discovered that another economic analyst had studied the Tasmanian economy recently and had cause to estimate multipliers. He kindly provided this analysis to us.

It is worth explaining, if ever so briefly, how the local economy operates. We wrote our survey questionnaire so to obtain the type of economic data we required to build our ground-up description of the economy. In fact, our questions to catchment residents were exactly as would be asked if we were to develop a regional input-output model based on a state-wide one. We needed to know what is bought where, at what cost, and likewise what is sold where, and for what price.

An input-output transactions table illustrates the flow of goods and services – and dollars – to and from the various industries and sectors. It helps us answer the question: what is the economic impact on the surrounding towns of both personal and business expenditure by catchment residents? Catchment residents' spending of money in Oatlands or Triabunna (to take two examples) goes to make these towns what they are. Their size in population numbers, the types of business that are viable in the towns, and the number of people employed in them are (in part) determined by the economic decisions made by catchment residents. Catchment residents can, for example, decide to purchase goods and services in Hobart, Sorell or locally in, say, Swansea. It is the case that some services can only be bought in the capital city Hobart. This is because that is where the businesses or government agencies exist. For example, a farmer in the catchment is likely to send a cheque annually to an insurance company in Hobart. On the other hand, the purchase of foodstuffs and a range of personal items is where considerable choice can be exercised.

The type of analysis we are discussing here is not just a simple exercise of counting the flow of dollars or the number of workers. Economic activity is like relationships in ecology – someone's output is another person's input, and the latter's output is someone else's input, and so on in ever dwindling quantities. These subsequent rounds of activity economists term 'flow-ons'. As already described, flow-ons can be estimated by using multipliers derived from input-output transaction tables.

For those wishing to practise the science of economic modelling, we offer the following comments. We would use input-output analysis to describe a set of relationships in, say, a geographical area as large as the Murray–Darling basin. We would first divide the basin into a series of large regions in which significant economic activity takes place, and in which there exist large towns (and possibly a city). These would be areas (and regional economies) many times larger than the Little Swanport catchment. It is unusual to analyse an area as small as our catchment in input-output terms. Of course, this does not mean that this is not feasible or desirable.

We start with the basic principles. An input-output transactions table is based on data in the business accounts of a firm or a farm, plus data gathered in household expenditure surveys. The manner in which individuals (farmers, shopkeepers, or whoever) and their accountants gather, store and manage financial data is not directly suitable for constructing an input-output table. In the first instance, an individual's accounts have to be rearranged to show the monetary value of all outputs of the business at their destination. The same with inputs (products such as fuel). In other words, we need to know where products (outputs) are sold and where business and household goods and services (inputs) are purchased. A typical farm output is a food crop that is destined for a processing plant (often far away) on its way to the consumer (final demand) in distant markets.

The individual data from a group of producers in the same business – a business can be narrowly defined as, say, 'wool growers', or broadly as 'farmers' – is aggregated in a grouping that forms an industry or, even more broadly, a sector. All the industries and sectors in the economy under analysis can be arranged in a matrix with the outputs of one industry being the inputs to another. For example, one of the sub-categories of a formal sector defined by government economists as 'Agriculture, Fishing and Forestry' (a large and disparate sector) could be 'wool production'. The output (from a supplying industry) becomes an input into the formally defined 'Manufacturing' sector (a sub-category of which would be 'wool milling'). A wool mill is a 'using industry' in the 'Intermediate Demand' quadrant. Intermediate Demand is the step before a final product (say, a wool suit) becomes a part of 'Final Demand'.

Some supplying industries send their output to both 'Intermediate Demand' and 'Final Demand'. For example, a fruit grower might send most of his or her produce to a cannery while selling a small amount directly to the public from a roadside stall. By introducing Final Demand we have already gone beyond the flow of goods and services between interdependent industries, of which there are many, as we have introduced households into the modelling.

Contributions to the local/regional economy

We now go to our case study survey data. As will be seen, the various communities in the catchment make a significant contribution to the regional economy, which (as noted) is centred on the adjacent towns of Oatlands, Swansea, Triabunna, Orford and Buckland. It is in these small towns that the catchment residents purchase many of the goods and services using the income they have earned from their work and production in the catchment. Let us consider this in some detail. As the various catchment communities have different needs and their spending differs, it is sensible to separate them for the following discussion. Before that, it will help to clarify a few matters pertinent to the accounting that follows.

Irregular purchases

Major items of farm machinery are irregular purchases. They are 'capital investments' in economic terms. An outlay between $50 000 and $100 000 is not unusual when a purchase of farm equipment is made. Most types of farm machinery have lengthy effective (and long economic) lives. In the years covered by our survey data, very few farmers purchased major items of farm machinery. For some the drought that prevailed throughout most of the latter half of the first decade of the twenty-first century led them to postpone expensive purchases. Fortunately, the majority of farmers did not require new or replacement equipment during this period. In any case, for our purpose of constructing a bottom-up description of the economy of the catchment, major machinery expenditure would be excluded and treated as capital investment.

Stock purchases are in a different category of (generally) irregular purchases by farmers. As with farm machinery they are additions to the capital investment in the farm. Over the survey period, farmers bought and sold stock. With very few exceptions, the purchases were not significant, with the drought a major factor influencing farmers who might otherwise have bought stock. On the other hand, there were some significant sales of stock in response to the drought.

Farmers tended to fall into three groups: those whose stock purchases were in the order of $10 000 or less per year; those who, due to the nature of their farming business or the acquisition of more land, purchased stock to the value of $50 000 to $150 000 for each year of the three-year period; and those who purchased no stock during the period. The exception to these categories is a very small number of large farms whose owners feed large numbers of animals, and both considerable stock and quantities of feed were purchased during the survey period. More detail on the impact of the drought on stock rates is presented later.

Building new fences or repairing existing fences is an irregular cost for farmers; however, on all farms some level of repairs is required throughout a normal year, bringing some regularity to this. Fencing can be a major item of expenditure. In our catchment, much of the fencing work is undertaken by the farmers themselves and hence the cost includes the purchase of materials plus the 'opportunity cost' of the farmers' labour. Opportunity cost is the dollar measure of what the farmer could earn doing other productive (paid or unpaid) work in the time taken up in fencing. For example, it is conceivable that a farmer spending a month on repairing fences might have found local government employment for this period. The forgone wages would be accounted as the cost of his or her work on the fence. On the other hand, there might not be any opportunity for paid work outside the farm and the farmer would value his or her labour in fencing at zero dollars.

Expenses not included

A major expense for businesses (including farms) that are not owned outright but are being purchased is the interest payable on bank overdrafts and loans. Paying back the principal and interest is – or can be – a daunting cash-flow problem for the borrower. The higher the level of debt the greater the risk to the business when poor seasons occur. The average level of debt for our catchment farmers, which was in the order of $200 000 for all but the very small farms, masks some extremely high debts.

From an economic accounting perspective we can exclude from our transactions table the payment of principal. This is because who owns a farm (where both the farmer and a bank/lender share equity in the business) should not affect the functioning of the business and its profitability. We also exclude interest payments as these are simply the transfer of some of the business profits earned by the farmer and passed on to those who lend funds. That is, a farm's profit before payment of interest is the correct measurement of its economic profitability.

Lease payments

Some farmers will lease land in the appropriate circumstances. In this case, the use of the land for, say, grazing is expected to provide sufficient income to the lessee to cover all his or her costs (including a wage or salary for that person), plus the amount paid to the lessor. In our survey we found an insignificant number of lease payments being made.

Farm workers

A major cost item for a large farm is the payment of permanent workers including, when employed, a manager. Wool properties employ workers at shearing and crutching times but these are contract (casual) workers and in economic analysis are converted to 'full-time equivalent' workers. Only the very large farms in the Little Swanport catchment engage permanent paid labour. A small number of farms are run by employed managers as the owners are not residents. A few farmers did employ workers – either throughout the year or on a needs basis – during the period covered by our survey. The range of payments was very wide and the average very low due to the vast number of farms operating without employed labour. It is not possible to report the actual dollar amounts of wage or salary payments made as the number of farms with significant wage bills are so few that confidentiality would be transgressed.

The lower catchment residential community

We now come to the significantly different lower catchment. The relatively small number of medium-to-large lower catchment farms are not included in the following discussion. The small 'village' communities in the lower catchment have specific demands on the economies of the local towns. The people being discussed here include owners of small residential blocks (as in Pontypool and Saltworks Road) and hobby farmers, orchardists, and tourism operators. The fact that many of the non-farming community are not permanent residents and that a significant number are retired or semi-retired residents means that their household expenditure is not high by normal Australian standards. It is further reduced by the fact that the average household occupancy in these villages is under two people.

Taking these factors into account, the average household expenditure – made *when landholders are in residence in the catchment* – is estimated to be $300 per household, or approximately $150 per person, per week. Of this amount, 40 per cent is spent in Hobart, 40 per cent in Triabunna, 12 per cent in Swansea, and the remainder in Sorrell and Orford in roughly equal populations. With a new supermarket in Orford, this small town's relative increase in importance was starting to be noticed over the period we were interviewing catchment residents.

As noted, a significant proportion of the lower catchment landholders are not permanent residents, rather they use their land and houses on the land for weekend and/or holiday retreats. Virtually all non-permanent landholders have their permanent home in Hobart. By converting the time these temporary residents and their families spend at their Little Swanport weekender-cum-holiday home to the equivalent of permanent residents, the economic impact of this category of residents can be estimated. Data gathered from these people show that on average three non-permanent landholders equates to one permanent resident in terms of their economic relationship with the local towns; that is, 100 temporary residents equals 33 permanent residents in terms of their expenditure in the surrounding towns.

Household living cost across the total catchment

The average per adult weekly personal expenditure (across the catchment) on usual household goods (food, alcoholic drinks, tobacco products, cleaning/washing products, health products and medicines, newspapers and magazines, and entertainment such as video hire) has been estimated at just below $150 ($149.60). This is virtually equivalent to the per person expenditure in the lower catchment residential sub-category – differing by only 40

Table 15.1 Expenditure by catchment residents on food and related household products: location and amount

Town	Percentage	Expenditure
Oatlands	35	$1 400 000
Hobart	31	$1 240 000
Triabunna	21	$840 000
Swansea	6	$240 000
Sorell	3	$120 000
Orford	3	$120 000
Total	**99**	**$3 960 000**

cents. If $150 appears low, it is important to recognise that many farmers have their own source of meat, vegetable gardens are not uncommon, and a household comprising three adults plus a teenager would be spending between $500 and $600 per week on the listed items – not a small amount.

Education expenses, children's clothing, sports items and travel costs are not included in the estimates we are dealing with, but they are included later in our estimate of overall household expenditure. Obviously the expenditure per child increases with the age of the child. Very small children add virtually nothing to the expenditure on the items we have included. A teenage child will have a noticeable impact on food and entertainment expenses. We only include the expenditure on teenage children (otherwise young adults). From our survey data this is estimated at two-thirds the expenditure by adults; that is, marginally less than $100 per week.

It is important to note that for all households in the catchment a number of significant items of expenditure are excluded from the amounts presented above. They include important ones such as clothing, insurance, car registration, telephone, medical insurance, visits to doctors/hospitals/health professionals, replacement of household appliances, and other items that are normally paid for by sending a cheque to the service provider in Hobart. The other major expenditure item not included is motor vehicle running costs (petrol or diesel fuel, and repairs/maintenance etc.). While this is a substantial cost item, and for most residents the purchases are made in the adjacent towns, the difficulty of separating private from business expenditure has caused us to include motor vehicle as a business cost. We come to that below.

The total amount spent on the selected items by the catchment community is arrived at by multiplying the number of permanent residents by the figure of $150, adjusting the number to account for permanent teenage residents, plus the expenditure by non-permanents (adult and teenage) who were not at their catchment residences on census night.

On the basis of the average expenditure on the selected items (as enumerated above), the catchment community spends just under $4 million per annum, $3 960 000. It is possible to allocate this expenditure to each of the adjacent towns as well as further afield (for example, in Hobart). The breakdown is shown in Table 15.1, and Table 15.2 is an itemisation of the categories of expenditure.

As stated previously, there are various normal household costs that are not included in the above. These items in total, averaged out over a three-year period, amount to $3 150 000 per year for the total catchment population. To avoid double-counting the expenditure by businesses (mainly farmers and oyster farmers) on motor vehicle costs (a major item) and

Table 15.2. Household items

Item	Percentage
Food	50.0
Entertainment	6.5
Alcohol	9.5
Tobacco	3.5
Washing/cleaning products	6.0
Pharmacy/health products	5.0
Newspapers and magazines	3.5
Electricity	8.5
Other	7.5
Total	**100.0**

other expenses that overlap private and business expenditure, the amount should be reduced to approximately $2 million, which can be counted as household living expenses. The remaining $1 150 000 is included with business expenditure. As what is being described here is expenditure on services (such as telephones), clothing and appliances, most of the expense occurs in Hobart. The overall household expenditure on living costs for the total catchment is the previously noted $3 960 000 plus $2 000 000 – $5 960 000 (or just under $6 million per year).

Farm expenditure by various categories

Table 15.3 contains a comprehensive list of normal farm expenditure categories. We have constructed from the data we gathered a typical middle-sized beef and wool farm. It is not an average farm – given a significant difference in farm sizes and types in the catchment, the average is potentially a misleading construct. We could describe farms at various sizes as 'representative' of that size (and type) of farm. This is what we have done with our 'middle-sized beef and wool' farm. Later on, we present a hypothetical farm representative of a large farm in the catchment.

While the items listed in Table 15.3 are all essential for the successful operation of a modern farm, some tend to be more obvious and regular cost items than others. For example, weekly or monthly fuel purchases, which are a substantial cost of farming, require a steady cash flow; so do quarterly payments for electricity and local government rates. On the other hand, numerous expenses are annual (for example, the purchase of wool packs before shearing) to be paid when the wool cheque is received. The data in the table are based on information provided by farmers, but the table does not include the most severe effects of the drought in the latter part of the first decade of the twenty-first century.

Farm costs are traditionally divided into fixed and variable costs. The former have to be met regardless of what happens on the farm; for example local government rates are due even if the farm produces nothing much of value. The same applies to interest payments on borrowed funds. The costs of various services (such as insurance, banking fees and the like) tend to be fixed. Clearly the costs of shearing, potato digging, and anything else based on quantities produced are variable costs. For example, if sheep numbers are reduced because of drought conditions, the shearing bill is likewise reduced.

Table 15.3. A typical middle-sized beef and wool farm: percentage expenditure by item

Cost item	Percentage of total
Agistment	(a)
Animal health/veterinary services	3
Contract work (e.g. shearing)	14 (b)
Dog feed	insignificant
Equipment hire	1
Fence repair/maintenance	5
Fertiliser	8
Financial services/insurance/registration	9
Fodder	5 (c)
Freight	3
Machinery maintenance/repairs	4
Miscellaneous	4
Pasture preparation and maintenance	2.5
Pest control	0.5
Seed purchases	3
Rates and taxes/water	3
Selling costs/commissions/ads	5
Stock purchases	7 (d)
Travel	3
Vehicle fuel, repairs and maintenance	17
Telephone	3
	100%

(a) Agistment cost could be considerable in years where it is necessary and affordable.
(b) Shearing costs would be a far greater percentage of the total if the farm was mainly producing wool, or wool and fat lambs. In the above example, sheep are less important than beef cattle.
(c) Fodder costs would be much higher in severe drought conditions – assuming the farmers could afford to purchase it.
(d) Stock purchases are very much a farm-by-farm decision, and they vary significantly depending on grazing conditions.

Table 15.4 summarises expenditure data for a hypothetical (representative) large farm. Note we are not discussing the largest farms, but rather a representative farm earning enough to pay $150 000 per year for the listed items plus meet other considerable expenditure items and earn a livable income for an average family. We have constructed the table using the data we obtained from all farms having expenditure at approximately this level. Not all expenditure items are included and hence the total of $150 000 shown is considerably less than the grand total would be. As with Table 15.3, the data are based on information provided prior to the worst effects of the severe drought.

We have selected major items of farm variable costs (excluding household purchases, which we have discussed previously). The items shown in the table account for just under 50 per cent of farm total costs. Major items excluded are stock purchases and machinery purchases and major repairs. These could be considered capital costs and account for over a third of total costs. All these expenditure items, plus the remainder, are included in Table 15.5.

Table 15.4. The expenditure by a hypothetical large farm: per year

	Expenditure ($)	Percentage of total
Animal health (drenching)	20 000	13.0
Electricity	4000	2.5
Fencing materials	15 000	10.0
Fertilisers	31 000	21.0
Finance (accounting)	4000	2.5
Insurance	5000	3.5
Petrol/diesel	19 500	13.0
Rates	5000	3.5
Shearing	25 000	16.5
Seeds	2000	1.5
Sprays	12 000	8.0
Stock feed	2000	1.5
Transport	4000	2.5
Wool packs	1500	1.0
TOTAL	**150 000**	**99.0**

The flow-on effects

Having established what farmers spend on their main inputs, we turn attention to the flow-on economic benefits of farming. These are benefits to businesses that either sell to, or buy from, farmers. We will concentrate on those selling to farms – what economists call the 'backwards linkages' – as much of the economic impact of the 'forwards linkages' (the processing of farm product) occurs outside the regional and state economy – even outside the Australian economy, as in China for wool. There are, as a consequence of farmer transactions, benefits to the owners and employees working in associated businesses in terms of profits and wages – and simply in employment. The type of associated businesses benefiting from farm expenditure depends on the nature of the goods or services bought or sold.

Certain types of payments are normally paid to 'head office' in the capital city (Hobart). Insurance, banking and government fees fall into this category. On the other hand, a variety of purchases are sourced as close to the farm as possible – for the obvious reason, to reduce transport costs. Fuel, stock feed, seeds, sprays and fertilisers will be purchased

Table 15.5. Totality of expenditure by catchment residents per year

Location	Amount ($)
Buckland	<100 000
Hobart	7 205 000
Oatlands	4 500 000
Within catchment	750 000
Triabunna	1 312 000
Swansea	409 000
Sorell	320 000
Orford	120 000
Elsewhere	2 344 000
Total	**16 960 000**

locally if they are available in the quantities and qualities the farmer wants. However, not all farmers will buy at the closest source, for a variety of reasons (for example, long-running relationship with a distant supplier might be favoured because of that relationship, or maybe preferable payment terms are offered). When we consider the actual farm expenditure patterns below for the Little Swanport catchment, it will be noticeable that factors other than local proximity are at play.

In calculating the flow-on effects of industries within our catchment, we combine all industries in the catchment. This overcomes any potential commercial-in-confidence issues that could arise through a separate categorisation by type of farm. The oyster farmers and the oyster nursery are included with the terrestrial farmers. They are all farmers. Given confidentiality requirements we cannot discuss the three oyster farmers as a separate category. A guide to the importance of the oyster industry is to think of it as being equivalent to three large terrestrial farms. We will discuss the oyster industry in a little more detail later.

Caveats

It is important to note that the expenditure data we present in the tables in this chapter represents a snapshot averaged over the three years 2003–04, 2004–05 and 2005–06. The data are not necessarily for a 'normal year' – whatever that concept implies – but neither is it an 'abnormal' situation we are describing. The data do not cover the extreme drought conditions that gradually developed in 2006 and became obvious in economic terms in the 2006–07 year. We use data collected for the drought years later (for another purpose). It would result in a significant distortion to use averages that included the severe reductions in income and increases in defensive expenditure of the drought years. As a consequence of excluding the drought years, substantial defensive expenditure by farmers is not included. This would include the purchase of stock feed in 2006–07 and beyond. 'Defensive expenditure' is an economists' term to describe paying money either to attempt to drought-proof a farm or to compensate for the effects of drought by purchasing fodder (and, if possible, purchasing water).

Various other changes have occurred since we gathered the data. For example, fuel prices kept on increasing; the price of grains, wool and meats have increased substantially; but so has the cost of fertilisers. The dynamics of markets require regular updates of economic analysis, although in the case of the Little Swanport catchment it would be appropriate to wait for the effects of the drought to be worked through before another analysis was conducted – and rebuilding after a severe drought can take years. Other than fuel price increases, the three oyster farms (oyster and spat producers) have not experienced any significant changes in the costs of inputs or the price received for product. These have not been adversely affected by the drought, as we show later.

Given the vagaries of nature and markets, we must be circumspect in making generalisations. For example, fertiliser application (which is shown as a major cost) varies significantly depending upon the types of crops grown (if any) and the nature of sheep and cattle runs, such as the extent of improved pasture in comparison to 'run country'. The data we gathered from individual farmers show that even in years where there is not a significant variation in climate and/or gross income, fertiliser applications can be variable. There are circumstances when a particular farm will need to apply out-of-the-ordinary quantities of fertiliser, notwithstanding external factors. In fact, examples of this were found in our survey data. As a conservative strategy we excluded out-of-the-ordinary expenditure.

The other thing to note about averages is the range of expenditure, from the farm spending the least to the one spending the most. This can be (and is in our survey data) a significant range. As we have a small number of large farms (spending considerable amounts on a mix of inputs), we find the average is being pulled up from what it would have been if only the smaller farms were represented. On the other hand, the large number of small farms that were each spending small amounts on most inputs pulls down the average.

Expenditure does not show a linear relationship with farm sizes, mainly because larger farms and businesses generally have easier access to financial resources; that is, they find it easier to borrow and spend. There are also differences between the amounts of some inputs used on the lower catchment farms compared with the upper catchment ones. For example, the lower catchment farms require far less application of sprays (weedicides etc.), as there is not as much weed infestation in this area as in parts of the upper catchment. It is important to note that not every farm or business has to purchase every item from the lists. A weed-free farm would not need to use herbicides. In good years, most (maybe all) farms would not need to purchase stock feed. Rainfall can vary greatly between properties – this was very evident in the data we gathered from individual farms – and hence productivity varies significantly, as does the need to compensate for lack of rain (such as by the purchase of stock feed). Then there is the human factor. Farming is a very individualistic enterprise. Differences in the quality and quantity of farmed products – and profits earned – are inevitably related to the skills and attitudes of the individual farmers. Farm management skills influence the amount spent on farm inputs, as much as the market value of outputs.

We aggregate the data on farm expenditure, the data on expenditure by the oyster growers, plus the data from across the catchment on personal living expenses (the latter estimated previously at $5 960 000) to show the aggregate economic impact that the catchment residents and businesses make on the local towns. This is presented in Table 15.5. Nearly $17 million flows out of the catchment into local towns, the capital, and further afield.

As discussed earlier, the totality of economic input does not cease with the expenditure in the local towns, Hobart or elsewhere. There is a flow-on (multiplier) effect. For every $100 of purchases in the local towns (Oatlands, Swansea, Triabunna, Orford and Buckland) by catchment residents, the local businesses (such as the stock and station agents) must expand output (from what it would have been if the catchment people had not gone to them). Following this initial impact, those supplying the stock and station agencies also have to expand output, and so on in an ever-dwindling chain. Multipliers can be calculated for output, income and employment to illustrate the impact, and the multipliers are then used to calculate the total impact. Doing these sums (applying the multipliers provided by Dr. Julian Morison), we estimate that in the order of 150 jobs exist in the adjacent towns as a result of the business they get from the catchment and, in addition, there are about 70 catchment-related jobs in Hobart. This means that – at a minimum – there are 220 jobs created by the economic activities (mainly farm production) in the catchment. This equates to approximately one job outside the catchment for every one job in it. We could use the income and output multipliers to the same effect, but believe the employment multipliers are the easiest to comprehend.

To our next task

We have described the regional economy the best we can with the data available from our survey, and given the fact that there are no longer formal input-output transaction tables for Tasmania (a major impediment to what should be a simple mathematical exercise) we have relied on the good will of another economist to calculate the total local impact of the economic activities in the catchment. What we have not done is delve into the use of resources, in particular water, in our catchment. This is our next issue.

16

The Little Swanport water accounts

Peter Daniels

Water is always in circulation, moved in its various forms between phases and between locations. (Leaman 2005, p. 10)

Introducing the framework

The task at hand for this chapter is to present an appropriate water accounting framework for the Little Swanport catchment and to populate this framework with data where available or where they are capable of being estimated from related or proxy statistics. This water accounting research has been moulded around the water cycle within Little Swanport catchment and is undertaken in order to understand and measure the flows and interactions of water in the catchment, extending to its estuary.

As discussed, water accounting is undergoing fervent developmental efforts in a diversity of institutions, and the state-of-the-art will evolve rapidly over the next decade. The framework we have developed is offered as a step to aid this process and to help address the more immediate short-term concerns and pressures for socially efficient decision-making about managing water resources. One of the main benefits from water accounts when they are capable of being matched against broader socio-economic information systems is their contribution to water budgeting. This means that, where all costs and benefits are included, water can be put to uses that have the best economic outcomes for society.

In the opening sections of this chapter we describe the water accounting framework. This framework is generic in that it outlines a template that integrates the best and most relevant parts of several water accounting systems that currently exist or are under development. While our model is a blend of several approaches, it builds upon the essential logic employed in the National Water Commission's (NWC) (2008) 'Australian Water Resources 2005' water cycle and balance assessment approach. However, the framework has been customised in as much as it focuses upon a subset of the most relevant features and activities that influence water use and demands in our study area. It is important to note that our water accounting framework is a work in progress, and, like most theoretical and applied activity in this relatively new but critical area, it is likely to evolve and improve rapidly.

Knowledge of the economic structure and nature of human land use of Little Swanport catchment is vital for understanding and describing the area's water cycle characteristics. The land use and economic characteristics of the catchment have been described in some detail in earlier chapters. As shown, the Little Swanport catchment is predominantly rural/agricultural in character with a significant unplanned 'natural' area. Forty-four per cent of the catchment is classed as 'grazing modified pasture' or 'grazing natural vegetation'; an additional 11 per cent is 'production forestry', while another 44 per cent is conserved as relatively natural areas or exists as 'residual native vegetation'. However, much of this natural area land has been affected by humans at some stage over the past 200 years (for example by logging and partial clearing).

Water accounts for the catchment must be guided by its agricultural and rural character. The agricultural land, and even substantial tracts of the residual native vegetation areas, involve human clearing of original forest and other vegetation; stock habitation; other land cover and terrain modification; abstraction, diversion and storage of surface water flows and (very limited in our case) irrigation; and release of water back to the ground and the air via evapotranspiration. It is important to note that not all of the economic activities in the catchment relate directly to water falling in the catchment. As a consequence of a human-made water storage, Hobbs Lagoon, there is some water 'flooding' into the catchment from the neighbouring Prosser River catchment. Where there are differing amounts of water reported for the catchment's water regime, these are a consequence of either including or excluding the water spilling in from Hobbs Lagoon. In other words, this import of water is not consistently recognised as such, and we are in a position of having to deal with this on an ad hoc basis when we rely on the findings of others.

The water cycle and water account framework – logic and structure

The basic structure of the 'natural' water cycle and human intervention in the cycle is depicted in Figure 16.1. This structure includes both environmental and economic dimensions of the water cycle.

Without humans, the natural water account would be comprised primarily of surface water runoff flows fed by precipitation. Precipitation enters into soil water and groundwater, and surface flows can be fed by groundwater baseflow back into streams and rivers. Major losses between precipitation and watercourse surface flows out of the catchment occur due to evaporation from the ground surface (and all other surfaces such as paved roads), as well as evaporation and transpiration (through vegetation) from soil water. Non-human animal life uses water taken up by plants (or other animals) and may divert surface water flows by drinking from streams and springs. These quantities are likely to be very small in the overall water cycle.

The basic water cycle equation for the natural water cycle is (from left to right in Figure 16.1):

$$\text{Precipitation} - \text{Groundwater seepage} - \text{Evapotranspiration baseflow} + \text{Groundwater} = \text{Surface runoff}$$

Beyond the precipitation inflows and outflows, groundwater can include cross-catchment flows that do not follow the terrain constraints of the catchment. While groundwater provides an important medium, taking and returning from catchment surface water, it can have independent sources and releases that go beyond and across the system boundary. There is also the sizeable storage and movement of water in soil to consider in developing a comprehensive model of water in a catchment. Furthermore, water also enters and leaves the system as water vapour, and there can be very substantial flows of (usually) salt water into the catchment boundary via the estuary. All these flows are ignored in the water

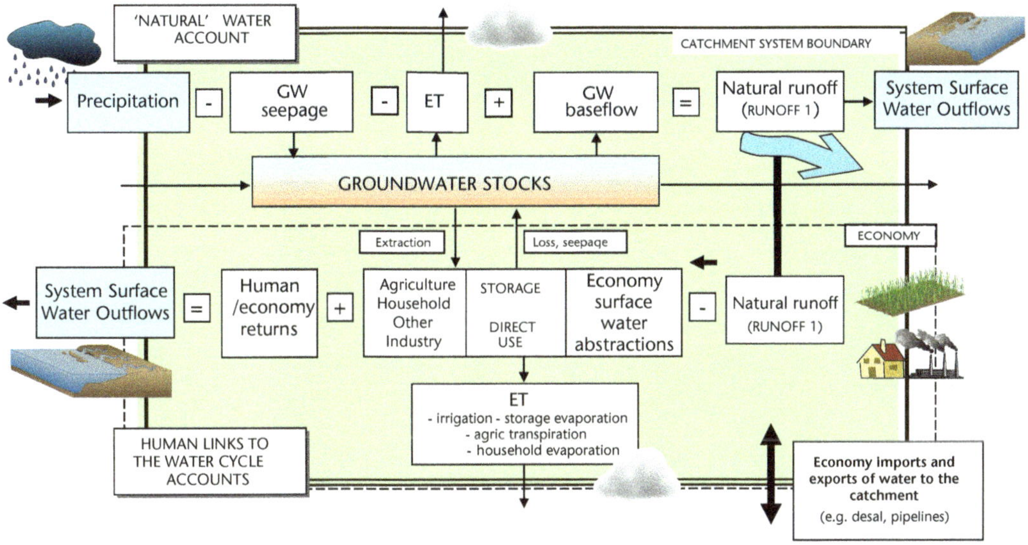

Figure 16.1. A simple water budget framework for catchment water accounts. GW = groundwater; ET = evapotranspiration. 'Natural' evapotranspiration and groundwater flows will be affected by human land use and extractions.

accounting framework presented here because for most of them there are no data and the cost of getting reliable data would be excessively high. While something is lost due to these gaps in knowledge, much is achieved by the modelling we can do.

When human activities are included (from right to left in the lower half of Figure 16.1), we have the abstractions from surface water that are returned (often with quality change of some degree) to surface water, or evaporated from impounded and irrigation surface water or from household use and release, or are transpired by agricultural biomass. Humans can build structures to import and export water from outside the catchment (for example the Hobbs Lagoon import on the boundary of the Little Swanport catchment in the south-west).

The simple water budget concept presented in Figure 16.1 illustrates the logic and structure used for the detailed water account framework described below (see Table 16.1). While our Little Swanport account draws heavily upon the approach adopted for the NWC's (2008) 'Australian Water Resources 2005' water balance assessments, it incorporates greater integration of economic and natural environmental aspects and provides different emphases in view of the specific characteristics of the catchment. These include the relatively high levels of 'natural' land cover (44 per cent); a particular economy and land use pattern and related water demands dominated by rural–agricultural activities; and a very small human population (521 permanent residents) and low levels of household and other water demands.

In line with the economy of the catchment, the Little Swanport account elaborates upon farm water uses such as farm dams and irrigation abstraction and loss, but has little detail on residential and other (industrial) components. The conceptual diagram in Figure 16.2 graphically identifies the major relevant water flows that are the focus of the detailed water accounts introduced in Table 16.1.

Notwithstanding the extensive data we gathered from farmers and other landholders during the many visits we made to the catchment, the measures of water use in what was a period of great variation in rainfall – from good rains to extreme drought – were clearly not

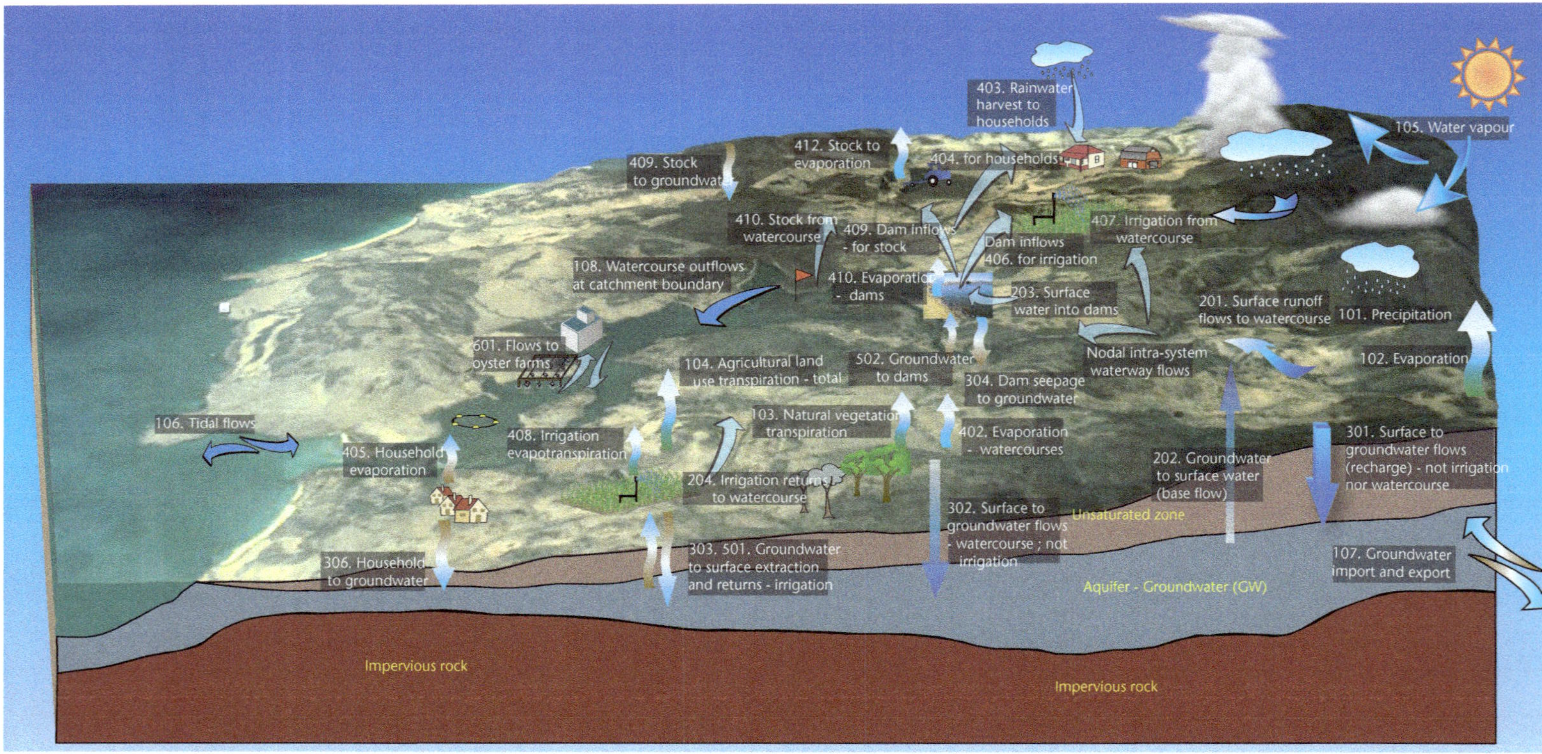

Figure 16.2 Little Swanport water cycle conceptual diagram. Codes match Table 16.1.

representative of more normal periods. Hence we were forced to rely on data from a much more stable climatic period. We are limited to using data from some years ago, 2002–03. There are no comprehensive figures for later years. For the purposes of consistency, all estimates are derived from sources as close as possible to the financial year we are using.

Mass water balancing

The water accounts applied to the Little Swanport context do take on some of the accounting features required for mass or material water balancing. This is essential for the long-term accuracy and usefulness of such information systems. However, inadequate data (especially regarding groundwater stocks and flows) have prevented the application of this logic to the available or derived statistics. In line with the mass balance approach, the Little Swanport catchment water accounts (LSCWA) begin with the system's opening storage balance for water. These stocks include natural and human stores of water. In our case study, farm dams are important but there is no accurate information on the level of actual storage at the accounting commencement date.

Unlike the Sinclair Knight Merz model used in the NWC's (2008) Australian Water Resources 2005 water balance assessments, the next section of the LSCWA identifies all the major inflows and outflows across the system boundary (that is, the catchment watershed). This includes precipitation into the catchment, evaporation and transpiration, water vapour movements, and flows from the sea and other sources outside the catchment. Separate measures are assessed for flows under human influence, such as evapotranspiration for agricultural land and irrigation, stock and domestic or household use, and the transport of water from outside the boundary. The next four sub-sections break down the main inflows and outflows between relevant components in surface water and groundwater.

Inflows to surface water incorporate flows from runoff and streams and rivers into storages. Both natural and human regulated surface water flows are covered. Human-related surface water transfers include surface water flows or extractions into farm dams and returns back from agriculture and domestic use to surface water. Tidal inflows can also be classed as surface water inflows but these are not measured here.

Inflows to groundwater comprise natural and human sources of water. Natural augmentation of groundwater results from pervious land surfaces and watercourses. It can also come from aquifer/groundwater movements into the catchment. Within the catchment, inflows to groundwater include those from seepage water storage on farms, from conveyance channels, and from household use.

Intra-system nodal flows have been added to the scheme to acknowledge the importance of the spatial distribution of water availability and use. Access to water is a major factor in determining effective availability. While there is very little information available on supply and use at disaggregated levels within the catchment, more specific flow (and stock) accounting measures that are geocoded and analysed within a spatial context will undoubtedly be essential aspects in future water modelling.

As we move into examining outflows from surface water, many flow measures from other parts of the framework that reappear on the list as outflows from one medium are often inflows into others (for example, flows into groundwater are often losses from surface water). The losses from surface water via natural means are via evaporation from watercourses and seepage to groundwater from water bodies (in the technical jargon, non-anthropogenic surface water features). A major loss from most catchments can have a human dimension. This is the flow of water from watercourses at the system endpoint boundary (often major water reservoirs). In the surface water outflows section of our model we find the most detailed analysis of abstractions and water returned to the economy as this is typically the primary source of water for human use.

Economic withdrawal of water (water withdrawn for agricultural production or other productive purposes) can be classified in many ways, such as self-extraction versus centralised or 'distributed' water, such as the mains supply in most Australian cities. This division will then be split across various industry types or sectors. In Little Swanport, there is no distributed water from centralised storage and treatment facilities. Hence, our accounts focus upon self-extracted water to be used by the main consumption sector, and from agricultural water use from farm dam storages. The surface water accounts track water transfers into dam storages. The outflow accounts measure irrigation, stock and domestic and household water use by surface water source (dams, direct watercourse extraction, rainwater harvesting and groundwater). As a subset of these uses, losses from the catchment system can be specified in the form of evaporation from dams and other built infrastructure. Where appropriate, accounts will differentiate between institutional types of water use (e.g. licensed or unlicensed). This is not a feature of the Little Swanport accounts.

A similar structure is used for outflows from groundwater. Nature-based outflows include discharge to watercourses, springs and evapotranspiration (the last is not included as part of overall evapotranspiration from the unsaturated soil zone). Economic extractions are primarily direct to dams for irrigation or stock and domestic use (and possibly household and other industry use). Groundwater use for economic purposes is not shown in the Little Swanport accounts, as neither we nor other researchers have been able to ascertain the extent of groundwater in the catchment.

Section 6 of the detailed water accounts in Table 16.1 shows the main economic sectors and water volumes by supply sources and losses and exports. It is primarily a reorganisation of measures from economic abstraction and returns in other parts of the accounts, grouping the information on sourcing and release by pastoral and cropping, household, forestry, and other economic activity. The closing balance is presented to preserve the ideal mass balance approach for the accounts, but, as with the opening balances, there are insufficient data to do so with the degree of accuracy we would like.

Preliminary water accounts for Little Swanport[1]

In the following section we put numbers to the water accounting framework presented above. As noted previously, the water account data is incomplete and is based on a mixture of existing data and estimates based on related or proxy data. Hence, it is important to stress the formative and provisional nature of the Little Swanport water accounts presented here.

A major source of good information on the Little Swanport water cycle is a study by Sinclair Knight Merz (2004). That study was carried out as support for the Tasmanian Government's Water Management Plan, and particularly to identify environmental flows and suitable entitlements in view of the availability of water resources within the catchment.

The rural and agricultural basis of the catchment has guided the detailed structure and focus of efforts in putting numbers in the water accounts. Agriculture is the major user of water in Australia and made up 65 per cent of human water consumption in the country in 2004–05 (Australian Bureau of Statistics 2006a). The proportion is slightly lower for Tasmanian agriculture, which comprises 59 per cent of the state's human water use. Overall, Tasmanian households consume about 16 per cent of the state's water, and manufacturing (primarily the wood and paper products industry) is responsible for around 11 per cent, with other uses being minimal.

For our Little Swanport accounts, urban and industrial water uses are insignificant or nonexistent. It would need a classification of the rural–residential areas of Saltworks Road and Pontypool as urban to make urban use a feature of the catchment. Our specific framework is suited to regions dominated by agricultural land use and farming. For example, there are few account items, and no data, for regulated or unregulated discharges that would be critical parts of the water cycle in an urban–industrial context. Notwithstanding the agricultural–rural focus of our accounts, the overarching structure that we have developed is based on inputs from a multidisciplinary team of researchers, and the detail and emphasis could readily be adapted to other regional economic structures where farming was less dominant.

First, if we consider just the overall, aggregated flows in the Little Swanport water cycle, the primary components are:[3]

- precipitation (500 GL)
- evaporation (213 GL)
- transpiration (155 GL)
- groundwater recharge (24 GL) and returns (6 GL) within the system (and ideally groundwater imports and exports through the system (unknown))
- surface water outflow from the system (approximately 113 GL).[4]

The simple annual water cycle equation, using the available data for the catchment, becomes:

$$\underset{\text{(Precip)}}{500\text{ GL}} - \underset{\text{(Evap)}}{213\text{ GL}} - \underset{\text{(Transpir)}}{154\text{ GL}} - \underset{\text{(Recharge to GW)}}{24\text{ GL}} + \underset{\text{(GW to baseflow)}}{6\text{ GL}} = \underset{\text{(Surface water outflow)}}{113\text{ GL}}$$

This can be conceived as a representation of the 'natural' water cycle though, as discussed, it is not possible to depict the natural regime clearly given that humans have markedly altered the catchment from the time of European settlement.[5]

The approximate precipitation level has been derived on the basis of a range of existing catchment rainfall estimates that yield between 360 and 565 GL. This equates to around 570 mm average annual rainfall across 880 km^2. Once again, it should be noted that drought conditions ensued after the accounting period we are using, but they are not incorporated as we are attempting to depict 'normal' conditions. The estimates for evaporation (at 43 per cent of rainfall) and transpiration (at 31 per cent of rainfall) are subject to uncertainty.[6] Based on the pattern of land use, around 50 per cent of transpiration (77 GL) is estimated from 'natural' land use areas and 50 per cent from agricultural and other human-influenced land. Catchment-wide inflows to groundwater are estimated as 3.5 per cent of total rainfall, with 25 per cent of this groundwater seepage returned to surface flows.[7] The balance (after evaporation and other losses) is considered as surface runoff into major watercourses and the average annual estimate utilised here is approximately 113 GL of surface water outflow into Little Swanport estuary (DPIWE 2003).

The detailed water accounts in Table 16.1 provide data, where available, at a more disaggregated level and focus upon the supply and use of water for economic (predominantly agricultural activity and farm dams) and household purposes. It is not necessary to discuss every individual data entry in the table. Only selected numbers are presented when they are of particular interest or relevance to the overriding aim of our research – the extent and significance of competing uses of water in the catchment. The most relevant data are summarised in section 6 of the table and discussed in the closing stages of this chapter.

The table is comprised of six sections and opening and closing stock balances. The six sections are:

1. Major inflows and outflows to the system (catchment)
2. Inflows to surface water
3. Inflows to groundwater

4. Outflows from surface water
5. Outflows from groundwater
6. Use of water, by supply source and release type.

Section 6 of the water accounts is primarily a reorganisation of other accounts in the table in order to facilitate the overall analysis of human intervention in the catchment water cycle and the relative importance and impacts of competing uses. The opening balance stocks have been discussed previously. However, we should once again emphasise that the estimates are subject to amendment given the lack of information about the capacity of natural reservoirs and human storages and the fact that there is no useful information about groundwater stocks. What is known is that the reservoirs in the catchment have been predominantly created by humans. In 2002–03, there were approximately 1200 non-flowing freshwater bodies in the agricultural part of the catchment, and the analysis by Sinclair Knight Merz (2004) suggests that total storage capacity was approximately 10 700 ML. This estimate is based on models that convert dam surface areas to volume. Non-licensed farm dams had a total capacity of around 6300 ML. There was 4100 ML of water in Hobbs Lagoon but, as noted previously, it acquires its water from the neighbouring Prosser catchment. There was an estimated 392 ML of licensed water entitlements in the Little Swanport catchment. Impediment structures divert flooding of water into adjacent areas of the Little Swanport catchment from Hobbs Lagoon. The importance of this flooding water is a moot point, but as it is not associated with extensive agriculture in the catchment we neglect its influence in our accounts. Human imports and exports of water across the system boundary are effectively zero, and there is no information available on cross-catchment groundwater flows. Only Hobbs Lagoon confounds what is a very simple catchment.

In section 1 of the accounts in the table, the major inflows and outflows to the catchment cover most of the major numbers used for the annual water cycle equation presented above. The only additional information in the table is:

1. the 3.45 GL loss/export of water from the system due to human water use from households, irrigation, dam surfaces and stock and domestic use (excluding that from evapotranspiration of precipitated water from agricultural and other modified land use)
2. the import and export of water vapour across the aerial system boundary (estimated at 4100 GL on the basis of Tasmanian data compiled by Dunlop (2001))
3. tidal flows into and out of the catchment (of seawater) are estimated to be 325 ML per day (or 119 GL per year).

Surface (fresh) water outflows are estimated, depending on the source adopted, as between 93 GL and 113 GL per annum.[7] This account item is a central one given the interest in the levels of economic extraction and use of water from the system, and potential water quantity and quality implications for the Little Swanport estuary and the oyster industry based there. The 113 GL surface water outflow estimate is adopted throughout most of the following analysis.

Most sections of the water account begin with indicators that are primarily 'nature'-based flows and finish with those linked more closely to economic activities. In section 2, the major sources of inflows to surface water are outlined. The level of runoff from precipitation represents water that enters major watercourses and ultimately the Little Swanport River at some stage in the catchment. It has been calculated as 119.2 GL by adding, to the 113 GL outflow in the estuary, 4.65 GL of net human extractions, a 1.5 GL net loss to groundwater from waterways, and 60 ML of water that would have been lost from the

Table 16.1. Preliminary Little Swanport catchment water accounts

Added to SKM – AWR (2005) model	Nature (N) or Human (H) influence	LS water account item code (see Fig 16.2) 'Item'	Account item	Estimate for Little Swanport catchment	Notes
			OPENING STORAGE BALANCE		
	N & H		Major on-river reservoirs (includes dead storage volume)		
	H		On-stream minor and farm dams (unknown, record as zero – assume same at start and end of period)		
	N & H		Major off-river storages (includes dead storage volumes)		
	H		Off stream minor catchment dams		
YES	H		Water entitlement water storages – licensed	4100 ML 392 ML (fed from Little Swanport catchment)	Most of this water is in Hobbs Lagoon and not drawn from Little Swanport catchment. Calculated from storage allocation for current entitlement (3846 ML) adjusted up given 90% demand use; 2002–03 (S1) * Based on current use; licensed withdrawal to dams reduces flows by only about 191 ML/year for stock and domestic use only (so relevant volume 191/0.5 = 392 ML). Entitlement basis = 689 ML/0.5 = 1378 ML
YES	H		Farm dams – non-licensed	6300 ML	Based on total capacity. Unknown % full (S1). Farm dam analysis * Around 1200 non-flowing freshwater bodies in agricultural part of catchment (in 1999–2000) (DPIWE 2003, p. 36) * Dams reduce annual flows by around 4760 ML

Added to SKM – AWR (2005) model	Nature (N) or Human (H) influence	LS water account item code (see Fig 16.2) 'Item'	Account item	Estimate for Little Swanport catchment	Notes
	N		Renewable non-saline groundwater (<3500 mg/L)		Fresh divertible groundwater stock – very crude estimate 8.6 GL as 22% of 39 GL (S4)
	N		Renewable saline groundwater (usually zero)		
	N		Nonrenewable groundwater (capable of being mined)		
			Total opening balance		
YES			**1. MAJOR INFLOWS AND OUTFLOWS TO THE SYSTEM (CATCHMENT)**		
YES	N	101	IMPORT Rain/snow/ice melt to surface	500 GL (Assumed from a range of 360–565 GL)	Estimates of the actual Little Swanport catchment area are variable: 610–898 km^2. 360 GL calculated as 610 km^2 (DPIWE 2003) × 591 mm (600–800 mm rainfall in other estimates) * BOM mean = 630 mm in 2002 (see RPDC 2003) but lower than normal since 1970; 630 mm × 898 km^2 = 565 GL Catchment area : 898 km^2 (LSCC 2002); probably includes outlets beyond main study * 880 km^2 from GIS land use map * perhaps should recalculate so is 491 mm/year as per water balance scenario for Little Swanport; = 660 km^2 × 491 mm = 294 GL
YES	N	102	EXPORT Evaporation – excludes transpiration; excludes small human use component	213 GL	* 43% of rainfall based on adjusted information in S2 – with 22% runoff vs. 10% * total evapotranspiration = 68% of rainfall based on S3; 31% + 43% = 74% based on S2 * Actual evapotranspiration is 500–600 mm per year (Source: BOM); not useful in this context

Added to SKM – AWR (2005) model	Nature (N) or Human (H) influence	LS water account item code (see Fig 16.2) 'Item'	Account item	Estimate for Little Swanport catchment	Notes
YES	N		EXPORT Transpiration	154 GL	* 31% of rainfall based on adjusted information from S2 with 22% runoff vs. 10% * covers natural vegetation and human modified land use
YES	N	103	Transpiration – natural vegetation	77 GL	Based on land use – 50% of land (S6) and 32% of rainfall evapotranspired (S2)
YES	H	104	Transpiration – from agricultural land total	77 GL	Based on land use - 50% of land (S6) and 32% of rainfall evapotranspired (S2)
			EXPORT Other human evaporation or transpiration - includes evaporation from households (405) + evapotranspiration and/or loss from irrigation (408) and stock and domestic water use (411) and from dams (413)	3.45 GL	
YES	N	105	IMPORT and EXPORT Water vapour	4100 GL	Not considered in water balance (net balance effect is zero). Largely independent of rainfall. Calculated as 880 km^2 * 4.7 GL/km^2 (S4)
			IMPORT and EXPORT Inflows and outflows from the sea to surface water		
		106	Tidal flows	119 GL	325 ML per day in and out; net = 0
			Desalination	0 GL	
		107	IMPORT and EXPORT Groundwater cross-catchment flows	?	

Added to SKM – AWR (2005) model	Nature (N) or Human (H) influence	LS water account item code (see Fig 16.2) 'Item'	Account item	Estimate for Little Swanport catchment	Notes
		108	EXPORT Surface water (watercourse) to sea	113 GL (Alternative: 93.6 GL)	* 113 GL (S7) Mouth of Little Swanport River; Source gives a minimum figure; gauge is upstream about 5 km from river outlet (Little Swanport and Ravensdale) into Little Swanport estuary; annual flows ranged from 12 GL to 190 GL; mean annual is 113 based on 1971–1990. * 93.6 GL (S1) * Seasonal flows are highly variable but winter flows are generally higher than summer; monthly flows also highly variable * note limited flow data in Little Swanport River based on only 19 years; did modelling to assess
			Unaccounted for rainfall loss	0 GL	Zero as has been calculated from runoff, evapotranspiration and deep drainage as percentage of rainfall (S3)
			2. INFLOWS TO SURFACE WATER		
	N		Rainfall to surface water runoff		Amount of runoff to surface water depends on many factors including land use, slope, storages and rainfall intensity, depth, spatial and temporal distribution
YES	N & H	201	Surface runoff flows into watercourses	119.21 GL i.e. approx 24% of rainfall (existing research estimates of runoff from rainfall vary from 7 to 45%)	Calculated as: 113 GL current outflow + 4.65 GL net human use withdrawal + 6 GL watercourse loss to GW - 4.5 GL GW discharge + 60 ML watercourse evap = 119.21 GL * Border Riv-G (2008) estimates 10% runoff * 7–10% runoff estimated for most land uses in Macquarie water cycle report NWC (2008) * 45% runoff for Tasmania (Bureau of Rural Sciences 2008)

Added to SKM – AWR (2005) model	Nature (N) or Human (H) influence	LS water account item code (see Fig 16.2) 'Item'	Account item	Estimate for Little Swanport catchment	Notes
	N	202	Discharge from groundwater to surface water (baseflow)	6 GL	Calculated from net deep drainage = 18 GL or 3.5% of total rainfall (S2) with groundwater to surface water baseflow being 25% of level of groundwater recharge from rainfall (S3)
YES	H	203	Surface water into dams	4.921 GL	Includes farm dams and relevant licensed dams (latter is only 191 ML) 3330 irrigation + 1430 stock and domestic – 30 ML to domestic households + 191 stock and domestic for licensed dams = 4.92 GL
YES	H		Licensed extractions to dams	191 ML	Is all for stock and domestic (S1)
YES	H	106	Inflows from sea – tidal flush	119 GL	
	H		Returns from economy inside entity (total of urban and irrigation where listed separately)		
	H		Urban treated effluent	0 ML	
			Water applied to irrigation and stock and domestic use	3.469 GL	Calculated as total surface water into dams (203) 4921 ML – 960 ML dam evapotranspiration (calculated) – 492 ML groundwater seepage from dams (10% of diversions) = 3469 ML
	H	204	Irrigation returns (to watercourse)	347 ML	10% of applied irrigation and stock and domestic water (LS account item code 204) (3.469 GL) returned to surface water (estimate)
	H		Surface inflow from other entities	0 ML	
	H		Desalination	0 ML	
	H		Returns from the economy outside catchment	0 ML	

Added to SKM – AWR (2005) model	Nature (N) or Human (H) influence	LS water account item code (see Fig 16.2) 'Item'	Account item	Estimate for Little Swanport catchment	Notes
			Unaccounted for inflow (error item)		
			Total inflow to surface water		
			3. INFLOWS TO GROUND WATER		
			Natural (predominantly)		
	N		Recharge to groundwater (excluding irrigation) – surface water recharge to groundwater	24 GL	Calculated from net deep drainage = 18 GL or 3.5% of total rainfall (S2) with groundwater to surface water baseflow being 25% of level of groundwater recharge from rainfall (S3). Deep drainage 3.75% of rainfall (S5)
YES	N	301	Non-watercourse recharge to groundwater	19.2 GL	Based on assumption that 80% of surface water to groundwater seepage is not from watercourses
YES	N	302	Watercourse recharge to groundwater (seepage from streams to groundwater)	4.8 GL	20% of surface water to groundwater seepage from watercourses
	N	107	Inflow from aquifers outside catchment (small; assume zero) (groundwater import)	?	
			Agriculture/Irrigation		
	H	303	Drainage to groundwater from irrigation	693 ML	Includes stock and domestic to groundwater. Assumed 20% of applied water for irrigation and stock and domestic use (Item 205) (3469 ML)
	H & N	304	Dams – seepage to GWPart of (seepage from surface water features (e.g. dams, wetlands, etc	492 ML	Estimate only. No data on seepage from dams available; other surface water losses to groundwater covered elsewhere. Loss calculated as 10% of total water inflows to dams (4.921 GL)

Added to SKM – AWR (2005) model	Nature (N) or Human (H) influence	LS water account item code (see Fig 16.2) 'Item'	Account item	Estimate for Little Swanport catchment	Notes
	H		Conveyance losses (seepage from channels)		Likely to be included in Item 304.
			Other human		
	H		Seepage from septic tanks		
	H	305	Groundwater inflow from household	16 ML	Assumed as 20% of total water consumed (80 ML) and discharged on-site
	H		Aquifer reinjection (e.g. aquifer storage and recovery)		
			Unaccounted for inflow (error item)		
			Total inflow to groundwater		
YES			INTRA-SYSTEM NODAL SURFACE WATER FLOWS		* 65% of Little Swanport outflow at Lower Reach (Swanston). (S7)
			4. OUTFLOWS FROM SURFACE WATER		
			Nature		
	N	401	EXPORT Evaporation from open water and wetlands (excluding major storages and minor catchment dams)	?	
YES	N	402	EXPORT Evaporation from watercourses	60 ML	Surface area 100 ha × 10 000 m^2 × 600 mm evaporation/year
		302	Seepage from streams to groundwater	4.8 GL	Equivalent inflow to groundwater; 20% of surface water to grundwater seepage from watercourses (S2, S3)
	N & H	108	EXPORT Surface water (watercourse) to sea	113 GL	See end of this section

Added to SKM – AWR (2005) model	Nature (N) or Human (H) influence	LS water account item code (see Fig 16.2) 'Item'	Account item	Estimate for Little Swanport catchment	Notes
	H		Extraction to the Economy inside catchment		
	H		Urban Diversions		
YES	H		Centralised DIST/MAINS	0 ML	
	H		Household, industry use etc	NIL	
YES	H		SELF-EXTRACTION		
	H		1. Households		
YES	H	403	Households – rainwater harvesting, stormwater	26–74 ML	* Population = 521 estimate from 2006 Census (July); 750 (Little Swanport Catchment Committee 2006); additional 150–200 in Pontypool and Saltworks Road during holiday period) * 26 ML based on 140 L/day; 74 based on 390 L/day Tasmanian average, but would be less due to non-reticulated conservation * Dunlop (2001) 69 kL per person per year = 190 L/day/person
	H		Households – direct surface water extraction	0 ML	
YES	H	404	Household use from dams	30 ML	Catchment household water use from other sources estimated at 37–50 ML per year. Use from dams assumed as similar additional total quantity.
			Total household extraction	80 ML	Estimated as 30 ML from stock and domestic dams (Item 404) + approximately 50 ML rainwater harvesting (Item 403)
		306	Flows to groundwater from household	16 ML	Assumed as 20% of total household water extraction (80 ML) and discharged on-site

Added to SKM – AWR (2005) model	Nature (N) or Human (H) influence	LS water account item code (see Fig 16.2) 'Item'	Account item	Estimate for Little Swanport catchment	Notes
		405	Household release to evaporation	64 ML	Assumed as 80% of total household water extraction (80 ML) and discharged on-site
	H		2. Irrigation diversions (crop, pasture)		
YES	H	406	Inflows to dams for irrigation – total	3330 ML	
YES	H		Inflows to farm dams (unlicensed) for irrigation	3330 ML	Calculated from farm dam number × size analysis data (S1)
YES	H		Inflows to licenced dams to irrigation	0 ML	No licensed irrigation other than Hobbs Lagoon (not fed by Little Swanport catchment) (S1)
YES	H	407	Watercourse direct to irrigation	0 ML	
YES	H		Total surface water to irrigation	3330 ML	
YES	H		Losses		
YES	H	408	Irrigation to evapotranspiration and other losses excluding return to GW	2429 ML	Includes losses from stock and domestic use (Item 411). Calculated as the balance of total water diverted for these purposes (Item 408) minus dam losses and groundwater and surface water returns from irrigation and stock and domestic use. Part of 104.
YES	H	304	Irrigation water return to GW	693 ML	Includes losses from stock and domestic use (Item 412). Assumed 20% of water applied for irrigation and stock and domestic use
YES	H		3. Stock and domestic diversions		
YES	H		Farm dams (unlicensed) to stock and domestic use	1428 ML	S1
YES	H		Licensed dams to stock and domestic use	191 ML	S1

Added to SKM – AWR (2005) model	Nature (N) or Human (H) influence	LS water account item code (see Fig 16.2) 'Item'	Account item	Estimate for Little Swanport catchment	Notes
YES	H	409	Total dams to stock and domestic use	1619 ML	
YES	H	410	Watercourse direct to stock and domestic use	0 ML	Excludes rainwater harvesting. Unknown for stock.
YES	H		Total surface water to stock and domestic use	1618 ML	
YES	H		Losses		
YES	H	411	Stock and domestic use to evapotranspiration and other losses excluding return to groundwater	2429 ML	Includes losses from stock and domestic use (Item 408). Calculated as the balance of total water diverted for these purposes (Item 408) minus dam losses and groundwater and surface water returns from irrigation and stock and domestic use. (Part of Item 104.)
YES	H	412	Stock and domestic water return to groundwater	693 ML	Includes losses from stock and domestic use (Item 304). Assumed 20% of water applied for irrigation and stock and domestic use
	H		Urban stormwater		Covered in household above
	H		Minor catchment dams		Covered in irrigation and stock and domestic use above; household above
	H		Private diversions (on-site residential greywater re-use)		
	H		Rainwater tanks		Covered in household inflow from surface water (Item 401)
	H		Environmental extractions (consumed within Basin)		
			System losses – storages		
	H		Evaporation from major storages		

Added to SKM – AWR (2005) model	Nature (N) or Human (H) influence	LS water account item code (see Fig 16.2) 'Item'	Account item	Estimate for Little Swanport catchment	Notes
YES	H	413	Evaporation from all dams	1000 ML	Includes farm dam surface area + Hobbs Lagoon and other licences
YES	H		Evaporation from farm dams – unlicensed	960 ML	1.6 mill sq m surface × 1000 L (per cu metre) × 0.6 m/year evaporation (derived from S1)
YES	H		Evaporation from entitlement, licensed dams	40 ML	Estimated in ratio of licensed dam storage water use to farm dam storage (4%)
	H		Losses from infrastructure and operational losses		
	H		Evaporation from channels		Included with assumed irrigation losses
	H		Losses from minor catchment dams		Included with evaporation losses from all dams above
	H		Conveyance losses (seepage from channels)		
	H & N	305	Seepage from surface water features (e.g. dams, wetlands, etc.) and other losses to groundwater (dams here only)	492 ML	Estimate. No data on seepage from dams available; other surface water losses to groundwater are covered elsewhere. Loss calculated as 10% of total water inflows to dams (4.921 GL)
	H		Flood plain harvesting		
	H		Treated effluent discharges	0 GL	
	H		Urban diversions	0 GL	
	H		Irrigation diversions		
	H		Unregulated flow (floods)		
	H		Releases from dams to satisfy environmental commitments downstream of entity	0 GL	
	H		Rainfall/runoff harvesting		
	H		Transfers to other entities	0 GL	

Added to SKM – AWR (2005) model	Nature (N) or Human (H) influence	LS water account item code (see Fig 16.2) 'Item'	Account item	Estimate for Little Swanport catchment	Notes
			Other flow out of entity		
YES		106	IMPORT and EXPORT Tidal inflows and outflows to sea	119 GL	325 ML per day in and out - net = 0 (see Chapter 9)
			Aquifer reinjection (e.g. aquifer storage and recovery)		
			Unaccounted for outflow (error item)		
			Total outflow from surface water		
55	H & N	108	EXPORT Total watercourse outflows at catchment (system) boundary	113 GL	See previous descriptions in Section 1.
			5. OUTFLOWS FROM GROUND WATER		There is little use of groundwater in Tasmania and no information on groundwater in catchment.
			Nature		
	N	202	Discharge from groundwater to surface water (baseflow)	6 GL	Calculated from net deep drainage = 18 GL or 3.5% of total rainfall (S2) with groundwater to surface water baseflow being 25% of level of groundwater recharge from rainfall (S3). Important during periods of low flow.
	N		Groundwater discharge to springs		
	N		Groundwater discharge to evapotranspiration (does not include unsaturated zone usage by vegetation)		
			Extraction to the economy inside entity		
	H		Urban diversions		

Added to SKM – AWR (2005) model	Nature (N) or Human (H) influence	LS water account item code (see Fig 16.2) 'Item'	Account item	Estimate for Little Swanport catchment	Notes
	H		Irrigation diversions		Irrigation and stock and domestic use accounts as for irrigation but no data available but groundwater use is minimal in Little Swanport catchment
	H		Groundwater extractions		
		501	Groundwater to irrigation	200–300 ML	Not included in farm water use calculations above. S10 (p. 10); 197 ML estimate from ABS (2006b); 220 ML as 22% of area of Southern Statistical Division (S4) Assumed as mainly irrigation.
x	H	502	Groundwater to dams	0 ML	Relatively small volumes (S8)
	H		Petroleum wells		
	H		Extraction for aquifer reinjection (e.g. aquifer storage and recovery)		
	N		Inter-aquifer outflow		
	N	107+D54	Aquifer flow out of catchment – groundwater outflow from system boundary	?	
	H		Extraction to economy outside catchment	0	
			Unaccounted outflows		
			Total groundwater outflows		
			6. USE OF WATER (by SUPPLY and RELEASE)		
	H		AGRICULTURE		
			Supply		
			Surface water diversions to dams – for irrigation	3330 ML	All from farm dams (non-licensed)

Added to SKM – AWR (2005) model	Nature (N) or Human (H) influence	LS water account item code (see Fig 16.2) 'Item'	Account item	Estimate for Little Swanport catchment	Notes
			Surface water diversions to dams – for stock and domestic use	1400 ML	1430 ML; 30 ML to domestic household use
			Groundwater – dams	0 ML	
			Surface water – direct	0 ML	
			Other supply sources	0 ML	Would include mains and from other industry in more diverse economy
			Release		
			Evaporation from dams (farm dams)	960 ML	
			Seepage from dams	492 ML	Estimate. No data on seepage from dams available; other surface water losses to groundwater covered elsewhere. Loss calculated as 10% of total water inflows to dams (6.4 GL)
			Returns to watercourse	347 ML	From irrigation
	H		HOUSEHOLD		
			Supply		
			Main reticulation	0 ML	
			Surface water – dams – domestic	30 ML	
			Groundwater – dams		
			Surface water – direct		
			Rainwater harvesting	50 ML (26–74 ML)	See surface water outflow entry above (though is stored as surface water within system)
			Other supply sources		
			Total household water use	80 ML	Assumed as 30 ML dam water use + 50 ML rainwater harvesting use
			Release		

Added to SKM – AWR (2005) model	Nature (N) or Human (H) influence	LS water account item code (see Fig 16.2) 'Item'	Account item	Estimate for Little Swanport catchment	Notes
			Evaporation of water/wastewater discharge from household	64 ML	Assumed as 80% of total water consumed and discharged on-site
			Return to groundwater	16 ML	Assumed as 20% of total water consumed and discharged on-site
			Returns to watercourse	0 ML	
	H		FORESTRY		No additional extracted or diverted water used for production forestry or plantations in Little Swanport.
	H		OTHER INDUSTRY		
YES		601	Oyster farm flows		Complex patterns of supply and use of fresh and saltwater for oyster and fish farms
			CLOSING STORAGE BALANCE		
			Closing balance – water in store		
	H & N		Major on-river reservoirs		
	H		On-stream minor and farm dams (unknown, record as zero – assume same at start and end of period)		
	H		Major off-river storages		
	H		Off-stream minor and farm dams (assume zero as accounted for under catchment farm dam diversions)		
	N		Renewable non-saline aquifers		
	N		Renewable non-saline groundwater outside groundwater management units		

Added to SKM – AWR (2005) model	Nature (N) or Human (H) influence	LS water account item code (see Fig 16.2) 'Item'	Account item	Estimate for Little Swanport catchment	Notes
	N		Renewable saline aquifers and groundwater (neglect – not part of water resource)		
	N		Nonrenewable aquifers and groundwater (capable of being mined)		
	N		Soil – unsaturated zone (assume no change – hence ignore)		
	N		Snowpack (no snow)		
	N		River channels (record as zero – assume same at start and end of year)		
			Unaccounted for storage (error item)		
			Total closing balance		

Sources (shown in notes column): S1 = Sinclair Knight Merz (2004); S2 = Border Rivers–Gwdyr Catchment Authority (2008); S3 = Australian Water Resources (2005) (NWC 2008); Macquarie Water Cycle report; S4 = Dunlop (2001); S5 = Bureau of Rural Sciences (2008); S6 = Little Swanport land use map GIS (Figure 13.1); S7 = DPIWE (2003); S8 = Resource Planning and Development Commission (RPDC) (2006); S9 = Meyer (*pers. comm.*); S10 = DPIWE (2005b) watercourse surfaces. The discharge from groundwater to surface water (baseflow) is, somewhat bravely, estimated as 6 GL from ratios indentified in other Australian settings; such is the best we can do until groundwater in the catchment is studied.

The surface water flow levels to dams are important and represent the major part of the human diversions of water for agriculture. An amount of 4.92 GL is calculated from the 2002–03 farm dam analysis by Sinclair Knight Merz (2004) and comprises 4.75 GL from non-licensed farm dams and 191 ML from licensed sources. When the additional estimate of approximately 80 ML of water abstracted by households is added to this flow into dams we obtain total diversions of surface water of close to 5 GL.[8] In terms of the net impact on likely watercourse flows, we have to deduct the 347 ML that is likely to return from water applied for irrigation. Hence, the overall economic extraction from surface water flows out of the system would be 4.653 GL.[9]

In section 3 of the water accounts, estimates of inflows to groundwater are presented. The estimates are open to revision once thorough groundwater investigations are undertaken. The 24 GL recharge from precipitation has been used in the annual water cycle equation above. The major sources of groundwater inflows from humans would come from irrigation water and dams (estimated as 693 ML and 492 ML per year respectively).

The remaining discussion focuses upon human intervention and competition for water between economic uses in Little Swanport catchment. The competition, if there is any of significance, is between upland catchment farming and existing oyster production. Section 5 (the outflows from groundwater) has two accounting item entries – the discharge from groundwater to surface water (6 GL) and an inflow for surface water with a speculative estimate of 200–300 ML of groundwater extracted for farm water use. This groundwater is modelled in the accounts.

The outflows from surface water have the most detailed data in the accounts, given better information and the clear relevance of outflows. Three major 'natural' losses from surface water are losses to the overall system – evaporation from open water and from wetlands and watercourses (the latter calculated as 60 ML) and, of course, the large volume of outflow from surface watercourses to the estuarine and oceanic system boundary (previously identified for our analysis as 113 GL per year).

There are no distributed or mains water flows in the study area as supplied in modern cities and towns. The self-extracted volumes to the economy have been split into irrigation, stock and domestic components, though evaporation and seepage losses to groundwater are combined for dam irrigation and stock and domestic use once the water has been taken from the storages. Losses from dam storages (960 ML per year for non-licensed dams) are presented separately following these estimates. We believe that most of the results, in particular ones not discussed in detail, are self-explanatory.

Human impacts on the Little Swanport water cycle

Detail from the economic side of the environmental–economic resource accounts has many potential applications. In our Little Swanport catchment, there are at least two purposes. First, we need to gauge the overall extent of human intervention and demands upon the catchment's water cycle. This gives an idea of the extent to which the natural flux of water is disrupted by human activities. Second, data about water use for different economic and household purposes (and their sources and releases or losses) are vital for assessing competing demands and to determine management options that will be the best for society overall. In a nutshell, given limited flows of water in the catchment, where does its human use lead to the 'best' outcome? For our framework, 'best' is measured in both broad economic and social terms. In view of a concern with impacts of catchment land uses on those in the estuary, a particular interest is to understand and measure the human impacts on surface water outflows in terms of interrelated water quantity and quality. The water demands for each major economic sector are presented and explained briefly below. This is followed by the summary and description of the most significant surface water interventions. Groundwater use is not considered in the following discussion.

Apart from agriculture, water use by other industry in the catchment is minimal and is unlikely to have a significant impact on surface water flows. This is because there is no significant industry in the catchment. There is also minimal supplementary application of water (that is, irrigation) for the 11.7 per cent of the catchment under plantation and production forestry. Of course, this could change in the future if plantation forestry expands and irrigation is deemed desirable in the early stages of tree growth. Economic use for forestry is relevant to the water cycle in terms of its impact on runoff and evapotranspiration (reducing runoff and immediate surface water flows). The fact that plantation forestry replaces cleared land, and the application of fertilisers to this land, makes it an important factor for change in the area.

Household water use has been based on an estimated permanent population of 521 people. There are no reliable statistics on residential water consumption per person. Rainwater tanks provide the water, and personal attitudes and habits determine usage. Householders keep an eye on the water in their rainwater tanks – particularly in times of severe drought – but have no more reliable method of establishing the rate of extraction than the traditional tapping down the outside of the tank. Proxies of domestic use range from 140 L/day/capita for consumption in drought-afflicted regions of Australia (notably south-east Queensland in the late part of the first decade of the 21st century) through to 390 L/day/capita (the Tasmanian average in 2006) and 197 L/day/capita for the Southern Statistical Division (Dunlop 2001). Hence, using the catchment's permanent population figure, we get a range from 37 to 84 ML per year for residential use. An estimate of 50 ML has been selected. However, given the rural or semi-rural nature of much of the residential population, an additional 30 ML has been allocated from the 'stock and domestic' extractions in farm dams. Overall, residential water use is therefore estimated at 80 ML per annum. As expected, agricultural water use takes the lion's share of human water use prior to estuary surface water outflows.

Two main approaches have been used to assess water extractions and demand by farmers. First, water use has been estimated on the basis of the number of dams of different sizes that have been identified in the catchment. The primary source of these data is the Sinclair Knight Merz (2004) report, which uses the TEDI model and the frequency distribution by size of identified farm dams. The estimation procedure is complex and involves conversion of surface areas to volumes and a mix of assumptions and evidence about the relations between type of use and annual capacity use. The Sinclair Knight Merz data has been used to split the total farm dam water use of 4.76 GL into an irrigation (3.33 GL) and stock and domestic component (1.43 GL). Note immediately that the irrigation component is not realistic in terms of catchment water. As discussed above, 30 ML of the latter withdrawal has been allocated to domestic or residential use. Based on a sub-catchment (2002–03) use survey, an additional 191 ML from licensed allocations has been added to the stock and domestic use levels. Water entitlement information is of limited use to the study, given it does not involve actual use and the vast majority of the allocations refer to Hobbs Lagoon. As we have noted, this is fed by water from outside the catchment. However, here we are relying on the Sinclair Knight Merz model and Hobbs Lagoon water is included. According to their farm dam based method, overall agricultural water withdrawals from surface water were approximately 4.921 GL (in 2002–03).

A second, and we believe more realistic approach, is to estimate agricultural water use by identifying the number of livestock and quantity of agricultural output (or land-use equivalents), and apply appropriate water use ratios to these levels. Based on 150 000 sheep in the catchment in a typical year[10] and drinking, service and water loss levels of around 7 L/day each (DPIWE 2003; Marwick 2002), direct stock needs can be calculated as :

$$1150\text{ ML} = 150\,000\text{ sheep} \times 21\text{ L/day} \times 365\text{ days}$$

If 1.15 GL is added to the 3.3 GL estimated for irrigation purposes, we obtain around 4.5 GL of agricultural water use – close to the outcome based on farm dam statistics. The sum of 1.15 GL is a figure we can feel confident about. It is important to note that in later chapters, where the very severe drought of the later part of the first decade of the twenty-first century is taken into account, the data used here is replaced with information based on drought conditions.[11]

In this alternative approach, there would be other uses of surface water that would add to this estimate. For example, there is potato growing and other horticultural production, although limited. The only available research suggests that the use of extracted water (from catchment flows) for cropping is minimal. Our research did not turn up evidence of significant use of water for crops. Finally, there are the water requirements in the estuary. Water requirements for aquacultural activities in the Little Swanport estuary are very complex and difficult to link to catchment water accounts. This matter is dealt with in another chapter.

Table 16.2 presents the main water account data, relevant for assessing economic surface water impacts prior to the estuary outflows. It covers agricultural and household-related flows only (the latter component is relatively insignificant). The items have all been introduced and described previously in this chapter and have been compiled into this table to facilitate the assessment of the likely overall human impact on surface water outflows and potential estuarine water quality and quantity.

Table 16.2 shows that around 4.9 GL of surface water is extracted for agricultural purposes. Most of this is captured as non-major watercourse flows from farm properties into dams. The ongoing flows linked to this extraction are depicted by the arrows in the table. Evaporation and seepage losses from the dams reduce this quantity to around 3.5 GL, which is applied for stock and irrigation. In turn, most of this water is lost as evapotranspiration (2.4 GL) and losses to groundwater (693 ML), with 347 ML estimated as returns to surface water. If we add approximately 80 ML extraction for households, there is a total economic extraction of around 5 GL of surface water in the study area. The evapotranspiration losses (totalling 3.45 GL) are definitely a loss from the catchment system boundary and, effectively, the groundwater losses (totalling 1201 ML) can be considered not to provide new sources of supply into available surface water (though additional research could throw new light on this aspect). The estimated surface water returns from irrigation (347 ML) are deducted from the total extractions to generate an estimate of 4653 ML of surface water net extractions from the Little Swanport catchment.

If the net extractions are added to the post-economy surface water outflows to the estuary, we can estimate the total natural outflows into the estuary at 117.65 GL per year (= 113 + 4.653 GL). Hence, the economic extractions lead to a 3.95 per cent reduction in surface water outflows that would occur without humans.[12] As discussed, the actual water cycle outcomes that would be observed without humans are very difficult to identify, especially in view of the extensive land use modifications (affecting runoff, groundwater seepage, evapotranspiration and water quality) that have been only partly accounted for in these calculations. If the 93.6 GL figure for existing annual outflow is utilised (as per the Sinclair Knight Merz (2004) analysis), the reduction increases to 4.7 per cent. If we use the

Table 16.2. Overall extraction and losses by economic sector

Sector	Surface water volume extracted	Losses from evaporation and transpiration	Losses to groundwater	Returns to surface water
Irrigation	**3330 ML** (Hobbs Lagoon)	2429 ML	693 ML	347 ML
Stock and domestic	**1400 ML** (1430 – 30 ML to households) **+191 ML** (licensed source extraction) TOTAL Irrigation and Stock = **4921** ML		subtract	TOTAL water applied to irrigation and stock = **3469 ML**
Dams	Covered in irrigation and stock and domestic above	subtract **960 ML**	**492 ML**	–
Household	**30** ML (from dams) **50** ML (rainwater harvesting) TOTAL = **80 ML**	**64 ML**	**16 ML**	
TOTAL	**5000 ML**			

Total flows diverted by dams (excluding households) = 4921 ML = 3330 ML (irrigation) + 1400 ML (stock; non-licensed) + 191 ML (stock; licensed)
Total water applied for irrigation and stock = 3469 ML = 4921 ML (dam diverted flows) – 960 ML (dam evaporation) – 492 ML (dam seepage)
Fate of water applied for irrigation and stock = 3469 ML: 347 ML to surface water (10%), 693 ML to groundwater (20%), 2429 ML evapotranspiration (balance)
Total surface water diversion = 5000 ML = 4921 ML (dam diverted) + 80 ML (household)
Returns to surface water from the economy = 347 ML (Irrigation and stock to groundwater)
Net consumption of surface water (including losses to groundwater) = 4653 ML(= 5000 ML – 347 ML); 3452 if losses to groundwater are not considered consumption

stock water use levels based on number of sheep (or sheep equivalent) to account for beef cattle by water needs (1.15 GL versus 1.4 GL), the reductions are 4.7 per cent and 4.5 per cent for the 113 GL and 98.6 GL outflow levels respectively.

17

The value of water

Tor Hundloe

Our catchment is different from those that supply towns and cities with water for residential purposes and those that supply industry (for example providing cooling for power plants). It is a simple catchment, meeting simple needs – farming, oyster growing, and residential needs for drinking and washing water. We have already explained the controversy that has arisen over the proposal to change water allocation in the catchment by constructing a large dam in a corner of the upper catchment. Rather than to go immediately to the issue of the value of water if switched between users, we take the effort to explain practical human–weather interactions. The reasons for this will become obvious.

We start with farming. Most farming in the developed countries is a business, yet in many important regards it is different from other businesses. The vagaries of nature do not have to be dealt with by most industries, while droughts and floods have fundamental influences, both desirable and undesirable, on farming. Most disasters that affect the supply of farm products are natural ones. They can occur quickly – a flood, a cyclone – or take a little more time to have an impact, as a drought does. The vagaries of markets, such as a change in consumer attitude that puts fax machine manufacturers out of business, are gradual, and responses can be planned. Industries other than farming (manufacturing and the service industries) are, in general, protected from natural disasters.[1]

Extreme weather events are unlikely to affect the production and supply of white-goods, motor vehicles, or oil production, to name three common commodities. Factory operators do not have to build into their day-to-day, season-to-season, or year-to-year planning the probability of a major loss of production due to floods or droughts. On the other hand, farmers must expect droughts and floods. The idea of one bad year in seven is the conventional wisdom imbued in children growing up on farms everywhere in the world.

The extent to which farmers build extreme weather events into their expectations of the future is a key determinant of the profit, or loss, they are likely to make. Expect next year to be as good as the last year – when last year had optimum rainfall in the growing season – and financial disaster awaits sometime in the future when the poor season follows the good season. The converse holds true. The eternal pessimist will believe the drought will never break and won't be ready for the much needed restocking and replanting when the rain comes.

Obviously, the wise farmer considers the long-term future in his or her planning. In looking to the future, there is limited value in looking to the immediate past. The longer the period over which averages are calculated, the more likely it is that a 'true' average will be estimated. Even with long-term averages, the future will be unknowable. We are always dealing with expected outcomes and expected values, not actual (knowable) ones. What we expect is based on our past experiences, the past experiences of those we converse with, the things we have read, and any other valued source. Because of this, expectations are not subjective; rather they are, in part, objective, being based on experience and reasoned thinking. The more information we bring to bear, the more objective our expected future.

We can – and should – assign probabilities to future events. A probability, such as a one-in-ten-year flood, is based on good science. The less we know about the matters that influence a possible event, or the poorer our records of historical events, the less firm our assigned probability of a future event. We have been improving our predictive ability with regard to the climate. All that good science is now being reworked and rethought as we attempt to come to grips with climate change. Climate change has added a great degree of uncertainty with regard to the likelihood of extreme events. The better our knowledge of what drives weather systems, the more accurate will be our assigned probabilities. However it is not just the prospect of climate change that is making the task of forecasting extremely difficult.

There are three major influences at play in the early twenty-first century: the prospect of significant climate change; the rapidly changing Asian economy; and the threat of global over-population. All are decreasing our predictive capability because they are inter-related, and this makes the forecaster's task more difficult. We know that the growth of the middle class in Asia will increase demand for red meat, we know the same middle class will have a much higher carbon footprint than the rural Asian of old, yet we don't know when the Asian middle class will adopt environmentally friendly technologies, or when their industrial economies will start to reduce their carbon footprints.

Let us not place all emphasis on weather as the compounding variable. The market (supply and demand) is subject to another set of vagaries. These are generally beyond the control of farmers as a group, and certainly are beyond the control of an individual farmer. Significant shifts in exchange rates can have a major effect on profits for products sold in a set currency (usually the US dollar). Some currencies – in some periods – tend to fluctuate widely for no apparent rational economic reason, and hence make economic planning very difficult. Currency speculation is the root cause of the problem. So serious is it that Nobel laureate economist James Tobin has argued for a tax on international speculative transactions for decades. Australia would be a major beneficiary of the cessation of currency speculation. Don't hold your breath waiting for it. The speculators are more politically powerful than the economic reformers. The volatility of the Australian dollar over the past quarter of a century – notwithstanding the country's long-standing robust economy and political maturity – is impossible to explain on rational grounds. In the late 1980s the Australian dollar approached US 90 cents, it fell back to hit a low around US 50 cents a few years ago, to reach over US 95 cents in late 2008 and drop back to 65 cents in early 2009 and then climb to parity (equal value) in late 2010. These dramatic changes can only be explained by the action of speculators. The so-called economic fundamentals (that is, real-world variables) do not change as fast, as frequently and as dramatically as these fluctuations. Australia can be a business person's or traveller's nightmare, and the asserted benefits of a floating exchange rate with speculation allowed warrant serious questioning.

There are other difficulties facing a business in planning for the future. Consumer tastes change. A classical example of this relevant to our case study is the substitution of synthetic fibres (nylon, polyester etc.) for wool. This became obvious in the 1970s and contributed to the reduced demand for wool fabrics. It could be about to turn around with the dramatic increase in price for the major source for synthetic fibres, fossil fuels. Even if their prices ease in the future, the long term prospect – as we approach what is called 'peak oil' – is that their prices will increase and thus must favour wool and the other natural fibres.

The fluctuations in wool prices, production and exports have been one of the most dynamic features of the Australian economy over the entire history of the nation. A little more than 100 years ago Australia was one of the richest countries of the world because of the strong demand and high price paid for fine wool by the UK clothing industries. The demand for the very good quality Australian merino wool remained buoyant throughout most of the twentieth century. During the Second World War massive stockpiles of wool were contained in specially built wool stores. After the war exports became possible again. Not long after, during the Korean War (1950–53), Australian wool sold for 'a pound a pound'. It was an excellent price, the highest ever. Australia 'rode on the sheep's back' until 1970. Since then prices have fluctuated widely, so have sheep numbers, and consequently wool production. Changing the level of production in line with price changes is rational. If the price of beef increases while that of wool decreases, the farmer who can convert to beef production will. Sheep numbers will decrease, and those of beef cattle increase. Official statistics prove the point. Not all farmers can make the obvious, sensible change, due to the nature of their land or lack of finance – if sheep become of little value how do you make enough money by selling them to purchase cattle? Being a 'first mover', anticipating change and responding to it before others, can make all the difference between profits and financial ruin.

Farmers tend to rely on one product, or at the best a combination of a very few products (say wool and fat lambs, or bananas and pawpaws), and are not able to spread their earning opportunities (and risks) like other businesses in which a decline in demand for one item is compensated by an increase in demand for another. Think of the average supermarket. It does not matter that canned soups go out of fashion. Customers will be purchasing a substitute from the same shop. A wool property cannot produce a petroleum-based fibre. It is with this background in mind that we can start to understand businesses, and farming in particular, in the Little Swanport catchment.

Our case study

When we had gathered as much data as possible from farmers in the Little Swanport catchment and searched among it for the relationships between a producer's income and water use that we were expecting, the results simply did not appear. There were no statistically valid relationships. Farm profitability showed no pattern according to water use. There were two problems, but first a comment on the nature of scientific inquiry.

As with traditional scientific empirical investigations, the researcher commences by using the conventional and least costly research tool, even though the investigator knows from the start that the tool might not show up a useful result. It is, then, on to the next (usually more costly) research method, and so forth until reliable results are uncovered or the investigation is discontinued. Most science is like this. Much of it leads to a dead end, something the lay person does not necessarily appreciate, as (like the lottery) only winners make the news. Come the dead end, someone tries something new. If the question being investigated is likely to have (life-threatening) consequences – such as the diagnosis of a disease – the search will continue until the expert (say, medical practitioner) has sufficient information to make a decision as to what to do. For example, in the assessment for cardiac disease, a patient is likely to be put on a treadmill and required to walk/run as the initial step. If no obvious signs of a cardiac complaint are found, the next diagnostic tool is tried. Eventually, if the patient is still complaining of chest pains but no results are found, an intrusive tool (an angiogram) will be used. Each test is more costly and more accurate than the one before.

What were the problems we faced with the Little Swanport investigation?

First, there simply was not a sufficient number of farms with similar characteristics to allow for the construction of a representative production function. Recall the mathematical formula for a production function we described earlier. There is a relatively small number of farms in the catchment – of the total population of 521 people many are not farmers, and the average farm has a number of household members. Then there is great variability in farm size and type of farming. There was nothing we could do about this except hope that a large number (the majority) had similar water-use demands and earned similar levels of income. This proved not to be the case. From our surveys and aerial photography we discovered how many water-holes a farmer had (there are approximately 1200 in the basin) and how much pumping equipment was employed. Furthermore, we obtained information relating to government permission to access water and the fees involved. However none of this helped us develop a reliable production function. A major roadblock to our research had been encountered. Each and every farm had its unique water demands and production function. It was not going to be possible to model water productivity across the farming community.

Second, most farmers had no comprehensive data on water used, and this only added to our woes. Their animals drank from the large number of artificial water-holes or from creeks and rivulets that ran through their properties. It is not possible to measure water consumed in this way. This is in contrast to, say, fertiliser applications. All farmers had records of the monetary value of the fertilisers purchased and used, and likewise for a large number of other farm inputs. Water, not being bought and sold in a market, went largely unrecorded. And of course water is not simply consumed via animals drinking it – the grasses, clovers and other stock feeds are themselves fed by water, in the main by rain.

This indirect demand for water – we won't call it 'virtual water' as this term tends to be confined to water extracted or diverted from a natural source – is very noticeable on farms. Country turning brown, and remaining brown season-to-season, indicates a lack of rain – a drought. Good rainfall means good grass, good profits. Rainfall data is something farmers gather and work with. The relationship between rainfall and income is measurable. Here was a solution to our analytical problem. Rainfall (not water gathered in dams or supplied by humans) and income are what we will use to estimate the value of more, or less, water to the terrestrial farmers.

There are, as they say, many ways to skin a cat, and we found one in the variations in rainwater. They provided the link between profitability of farming and water, the very relationship that we were searching for. While this worked for the terrestrial farmers, the value of water (in both quantitative and qualitative terms) for the oyster farmers had to be addressed by a different approach, as we have seen.

The relationship between an environmental state (rain or no rain) and productivity (measured as monetary returns) is called the 'damage cost' when the state (quantity and/or quality) of the former reduces the latter. Damage cost analyses are one of the many techniques that have been developed over the past 40 years to ascertain economic values for environmental (or natural) resources. This technique is common to two sub-fields of economics, natural resource economics and environmental economics. Much of the early theoretical and practical work in popularising new economic tools was done by a group of practitioners at the East-West Center at Honolulu, Hawaii.[2] In 1983, their work resulted in the publication titled *Environment, Natural Systems and Development: An Economic Guide* by Hufschmidt *et al.* Two subsequent publications, *Economic Valuation Techniques for the Environment* edited by John A. Dixon and Maynard M. Hufschmidt (1986), and *The Application of Economic Techniques in Environmental Impact Assessment* edited by David James

(1994) contain numerous case studies, each utilising a different technique. In the last book, Hundloe uses a 'damage cost' approach to analyse a change in farming practices in the upper catchment of the Chao Phraya River, which drains much of Thailand. The river flows through the centre of the country and the region is known as the Chao Phraya River Basin, a fertile alluvial flood plain from which rice is produced and exported to other parts of Asia. The approach used by Hundloe is not dissimilar to what we used to obtain a measure of (increased/decreased) water to Little Swanport catchment farmers. Jeff Ross of our team estimated the value of water flows to the end-of-the-catchment oyster farmers.

The impact of drought in the Little Swanport catchment

The point of commencement is to seek correlations between rainfall and farm income, and note how farmers respond to changed weather conditions. Droughts are normal – and expected – occurrences in Australia. The concept of a drought is a relative term. A drought in a tropical rainforest area is vastly different from a drought in a semi-arid desert. The small Australian state of Tasmania provides a good illustration of this variability. But be ever mindful of the 100-year drought! The World Heritage-listed temperate rainforests of the south-west of Tasmania receive abundant rain – that is why they are called 'rainforests', while the east coast, particularly our case study area, is the driest in the state. A drought in the Little Swanport catchment area is likely to cause serious problems for farmers who, even in the best rainfall years, rely on the large number of small farm dams to water their stock.

The effects of an extremely severe drought started to be felt in the catchment in late 2006 and early 2007. There were drought conditions leading up to this period, but not as dramatic as they were to become. The previous year, 2005, had seen very good rainfall – more than normal. The severity of the drought varied across the catchment and hence its impact on farmers varied according to their geographic location and the type of country they farmed. Various environmental conditions, the extent of forest cover, soil type, and the duration of rainfall events (when they occur) have an impact on the value of water to a farmer. Put in other words, it becomes much more valuable when applied to good soils than to poor soils, when it falls in the growing season rather than at other times, when it falls precisely where crops and stock need it. From the onset of the drought, some farmers undertook a level of preventative or defensive expenditure; however, the worst impacts were not felt until considerably later, and only then did the costs become high.

Preventative (defensive) expenditure

Where a producer (or consumer) takes action to avoid a cost such as the loss of sheep or cattle in a drought, economists use the term 'preventative expenditure'. This assumes the action comes with a cost, which is usually the case. Money that would otherwise not be spent is outlaid to prevent something undesirable happening. A typical example of preventative expenditure is a household installing double-glazed windows if the house is in a flight path. The noise is avoided, but at a cost! Drought-affected animals can be saved, but at a cost of fodder purchases.

A loss of farm output is given the simple technical term in economics of 'loss of production'. Alternatively it is called 'forgone production'. The loss of farm output is 'damage' to the enterprise resulting from unusual environmental conditions. Damage tends to be caused by pollution and degradation of the natural environment; however, the same principle applies to any change that results in a less than optimal state. Measuring the loss of production is the favoured technique to calculate environmental costs. Its equivalent,

Table 17.1. Preventative expenditure leading up to 2006–2007

Action taken	Percentage of farmers undertaking this strategy	Average additional costs for those taking action ($)
1. Purchase more than normal stock feed	57	7600
2. Grow more than normal quantity of stock feed	43	20 000
3. Clearing-out existing dams, installing water tanks & troughs, improving irrigation, digging new water holes	36	15 000
Other (e.g. use more fertiliser, open up new paddocks for grazing)	14	Negligible

when workers' productivity declines because of environmentally unhealthy working conditions, is given the similar simple description 'loss of wages' or 'forgone wages'. Both types of losses are known by the generic name of 'damage costs'. Something unwanted happens to the environment (pollution, a flood, a drought), and damage is done to the productive capacity of the business. A flood through a brewery located on a river bank is a good example. This actually happened in the Queensland town of Rockhampton many years ago. Yet, in this case locals were known to remark that with some river mud in the pipes the beer tasted better. The exception that proves the rule, possibly!

The impacts of the severe drought on the Tasmanian east coast were not felt in the three years covered by our initial survey (2003–04, 2004–05, 2005–06). As a consequence, the farmers did not report spending much to counter the negative effects of the drought. Only by revisiting catchment farmers in late 2007 did we obtain data on the effects of what had by then become severe water shortages for many farmers. By 2006 all farmers were feeling some degree of distress.

The data in Table 17.1 indicates the action farmers were undertaking in the 2005–06 year to maintain productivity. We must emphasise the point made above that this was expenditure made before the severe effects of the drought were felt. The adverse impacts were still minimal. Some farmers undertook more than one strategy.

The cost to the farmer included the purchase of materials, plus the opportunity cost of his/her labour. The table makes it clear that even in this lead-up to the more severe drought conditions which were to follow, a sizeable percentage of farms were making preventative expenditure. The farmers' efforts and monetary outlays were to prove insufficient given what was about to happen. They had no idea of the severity of the drought to come.

The data in Table 17.1 indicate that many farmers were not taking steps to counter the drought or were not taking all options at this stage. This was for a variety of reasons, the most important being the marked differences across the catchment in rainfall and water availability (that is, the lack of water was not evenly felt), the shortage of funds (particularly for owners of small farms and those with high levels of debt), and differences in farmers' attitudes and expectations about future weather conditions. Any one of these reasons was justified by the circumstances of farmers in question.

Even when the extent of the drought became all too obvious it was not experienced universally in the catchment. Not every farmer faced the same dire straits, and not every farmer responded in the same fashion. The idea of an average farmer, making average responses to drought, is not a real-world concept. Don't expect any individual farm to have lost the same number of dollars as the average. Only by coincidence will there be an average farm. Some will have lost more, maybe much more; some will have lost a little, maybe nothing. However, based on our interviews with farmers and the financial data we gathered from them, it is possible to arrive at a range of responses that we can use to draw reliable conclusions. From these responses (in terms of preventative and/or replacement expenditures) we can calculate the value of changes in water quantity.

Before going to the data, it helps to spell out the choices faced by farmers.

A drought-affected farmer is faced with two major choices if the farm business is based on stock: sell down the stock, or purchase feed. In other parts of the nation there is a third option, and that is to purchase water from others who have water allocations in the same system. This is possible in parts of the extremely large Murray–Darling system, but even there is of limited scope because of the very 'thin' and immature water market. There was no one to purchase water from in the Little Swanport catchment when the drought became so severe that farmers would have entertained the idea had it been possible.

Both of the feasible responses (sell stock or purchase feed) follow from the fact that in drought conditions the carrying capacity of a farm is reduced. Reduce the flock or herd! Selling stock before they are ready for the market will result in lower than normal prices and obviously a drop in farm gross income and profits. Of course, if a farmer sells a lot of animals because of a drought, he or she is likely to experience a one-off increase in farm income in that particular year (notwithstanding the poor price for the stock). This is something to be on the lookout for in analysing farm profit and loss statements. The 2007 drought resulted in stock prices for Little Swanport farms falling on average by more than half (58 per cent) of their normal price. This was a dramatic loss for those who had to sell.

If there is no change in demand for beef or lamb/mutton, price will fall for one of two reasons. First, there is the excess supply at the going price – a significant number of sheep or cattle coming on to the market at the same time at a local saleyard can 'unsettle' the local market. Given that the number of animals put on the market in Tasmania (let alone that small part of it that is Little Swanport) is insignificant in terms of the Australian total, we can't blame an over-supply for driving stock prices down. The second reason for a low price is the change in the quality of the animal – one affected by drought is not going to be of top quality and will sell for less than the price it would fetch in optimal condition.

In the Little Swanport catchment, farmers who attempted to retain their total stock numbers, or at least more stock than could be fed in their paddocks, purchased fodder as the drought became worse. For some farmers this meant borrowing money. This was in a period when fodder prices were increasing (in response to increased demand elsewhere in Australia, as a consequence of the widespread drought) and the costs of borrowing were also increasing. The threat of inflation was recognised in early 2007 and the Reserve Bank commenced an ongoing program of increasing the formal interest rate that flowed through to the bank lending (or overdraft) rate.

The short-term impacts of severe drought can be difficult enough to cope with without meeting the costs of rebuilding the carrying capacity of the farm to its 'normal' state once the drought is over. Pasture has to be regenerated after a drought. If this is not done there will be permanent reduction in stocking rates, and hence in farm income. This type of expenditure is what economists call 'replacement cost'. In its general form, when a part of an enterprise is destroyed (say a farm bridge in a flood) and it has to be replaced before the enterprise can be as profitable as it normally is, the cost of replacing the damaged object indicates its value to the enterprise. The value of good quality soil to a farmer is the cost of replenishing it with fertilisers once the soil has been degraded by over-grazing or over-use. Replenishing eroded and degraded soil as a result of poor farm practices is probably the best example world-wide of replacement costs. Being in the financial position to re-stock immediately after the drought is a prerequisite for long-term profitable farming; however, this requires planning and in particular taking a long-term perspective.

Let us summarise at this stage. We have identified three techniques to calculate the value of extra units of water to farmers. They are the loss of productivity, preventative expenditure, and replacement costs. All these can be determined using market data – such as the price paid for fertilisers, purchasing feed, and rebuilding degraded soils. Economists would say that farmers (or whoever) 'reveal' through buying and selling decisions what the costs (benefits) are. No fancy surrogate or hypothetical markets have been brought into play when applying these techniques adopted from standard economics. The use of market prices means we can trust the data rather than be concerned about the biases that can result from less direct methods of assessing costs.[3]

The only other thing those in the situation of our Little Swanport catchment farmers can do is utilise existing water better, say by installing reticulation systems. The possibility of doing this is limited to the extent of water in the farmers' dams. In severe drought conditions this is not a realistic option. Whatever farmers do costs money that many haven't got.

Let us consider in more detail the farmers' responses to the changed environmental conditions they experienced. We will come to use one or other of the measurement techniques identified above to answer the question: what is the value of an additional unit of water to a terrestrial farmer?

Destocking

If de-stocking merino sheep, the first to be sold are wethers and dry ewes, for the obvious reason that they are not reproducing. If cattle need to be sold, the prize bull is last, the older steers first. They don't rebuild the flock or herd. In the Little Swanport catchment during the 2007 calendar year, wethers and ewes had been reduced to less than half (46 per cent) of the normal carrying capacity. This is an across-the-board reduction, with some farmers reducing by much larger percentages while others reduced by small numbers (including those able to maintain their complete flock).

Farmers will attempt to retain as many of the breeding ewes and cows as possible. However, during droughts the condition of these animals declines and lambing and calving percentages can be dramatically reduced. For example in the Little Swanport catchment, the average lambing rate declined in the 2007 drought from round 90 per cent to about 65 per cent. The reduction in the lambing rate is a simple measure of the stress on the animals.

The average lambing rate calculated over a period of years tends to become the expected lambing rate in farmers' minds. A farmer's planning (including financial decisions) would be made on the basis of an expected lambing rate of 90 per cent. A cattle breeder would have an equivalent calving rate in mind. Unfortunately there is a tendency to exclude from the estimation of the average lambing or calving rate the marked difference an extreme weather event (such as a drought or flood) will have on the average. When an extreme event occurs, as in the Little Swanport catchment in 2007, the differences between its expected normal and what occurs can be a serious loss of income to the farmer.

Using sheep as the example, the drought-induced reduced lambing rate meant that instead of 900 lambs for every 1000 ewes, there were only 650 lambs. This is a loss of 250 future wool producers (assuming we are dealing with merinos, not crossbred fat lambs). The same logic, but in terms of meat production, applies to fat lambs; however, the annual life-time of wool production makes for a significant difference in the total loss.

The wool loss is the net present value of the year-in, year-out loss of wool production over the wool-producing lifetime of the sheep, plus the sheep's final sales value, plus the lambs not born if the sheep is a ewe. The forgone wool production of the 250 sheep not born but expected to be born in a normal season would equal the weight of the fleece multiplied by the selling price (based on its micron count) multiplied by the number of years the sheep is shorn, minus the costs of maintaining and shearing the sheep. A number of simple arithmetic calculations are involved here, but all quantities are capable of reasonably accurate measurement.

As with sheep, drought conditions affect the number of beef cattle that a property can carry without importing fodder. The 2007 drought[4] resulted in a fairly dramatic reduction in breeding cows. The average herd in the Little Swanport catchment in late 2007 was approaching half of what it was a year previously (at 55 per cent). The calving rate also declined significantly. The same type of calculations as for sheep can be undertaken to measure the loss of beef production.

Feeding stock

Little Swanport farmers who could afford to purchase feed (either through the use of their own funds or by borrowing) did so. Our survey results showed that the extent of feeding varied significantly, depending on what level of stocking a farmer wished to maintain (or could afford to maintain), and on the condition of both pastures and run country on the property. Obviously the individual farm data that we were provided with cannot be reported for confidentiality reasons; however, we can say that the largest properties each spent on average, from the time the drought began (in the first half of 2006) to the second half of 2007, a period of approximately 18 months, $0.5 million or more on stock feed. This very high level of expenditure is an indication of the order of magnitude of the monetary value of farm output on the large farms. One does not spend this amount of money if the potential loss is not comparable. The smaller properties, which are the significant majority, spent about one-tenth of this on fodder during the same period. However spending is not necessarily commensurate with farm size, but rather with the ability of the farmer to source funds. Many farmers with small properties could not access the necessary money to buy any feed for their stock. Their plight was the worst.

Re-sowing pasture

One of the impacts of prolonged drought is the failure of crops. These include animal food crops as well as grains or whatever produce is destined for the market. For farmers running cattle or sheep, pastures are important. In the expectation – or hope – of the drought breaking, farmers re-sow pasture. This is a costly undertaking at approximately $400 per ha in our study area at the present time. One hundred ha of sown pasture would cost $40 000, equal to the total gross farm income of the very small farms.

Some simple sums

With this background, we can commence to put a value on water to farmers in our catchment. We will use the loss of productivity approach. We will come to an overall estimate later, but first it will be instructive to illustrate what we are doing by drawing on our survey data. We will construct hypothetical farms, and the value of units of water to them, so as not to identify any particular farmer and his or her situation.

A large property of 3000 ha would, in a normal year, run 8000 sheep plus 50 beef cattle. This is a hypothetical property in the Little Swanport catchment. Any resemblance to an existing property is purely coincidental. The sheep would produce 240 bales of wool, at 33 fleeces per bale. The average selling price for a bale of wool in 2007 was in the order of $1700.

The 2007 drought resulted in the sheep numbers being reduced on this property to 4800 sheep and the cattle to zero. This means there was in the order of 1.6 sheep/ha due to drought conditions, whereas in a normal year the same property would run close to 3 sheep/ha if no cattle were grazed. This reduced number of sheep would produce only 130 bales of wool. The average fleece decreases in weight as a consequence of the drought. The loss from the wool clip is 110 bales (240 minus 130), that is a loss of $187 000. This is the gross monetary loss from reduced wool sales. However, there are 3200 fewer sheep to shear, mules, muster and generally maintain, and hence less overall cost to the farmer. The net return per sheep is in the order of $30 (half the selling price of the wool). This means that the loss to this farm would be in the order of half of $187 000 per year, being $93 500. *This is the very minimum value of the rain that did not fall in the drought years.* We will present rainfall data later so that income losses can be compared to decreased rainfall. The above does not include the income earned in the year for the sale of the sheep. Only a small number of the 3200 sheep would have died in the farmer's paddock; most would have been sold at discounted prices as discussed above. These sales would add to the farmer's income in the year of the sale but (as explained previously) reduce the farmer's income in future years.

If the cattle on this property were sold the income would also show up in the year of the sale and the loss (in terms of calves) would be in future years. As both the sheep flock and cattle herd have been reduced, the value of the property would be coincidentally reduced.[5] The value of a property is largely determined by its profitability. In strict economic terms it is the 'net present value' of its profits year in, year out. The two tend to change in tandem.

For the hypothetical farm we have been discussing, the value of the water that did not fall in the period can be estimated by the net forgone production, which is 3200 sheep at approximately $30 per sheep per year. This is the minimum loss, as only forgone wool revenue is counted. For simplicity's sake the cattle are left out of the arithmetic, as are losses in future years as the flock is rebuilt.

Another large (hypothetical) property (of 3350 ha) would in a normal year run 11 000 sheep plus 200 beef cattle. You will note that this property, although just over 10 per cent larger than the previous one, can run significantly more sheep and cattle. It has more better-quality grazing land, more improved pastures, and a farmer more attuned to the business of farming rather than the lifestyle of farming. We are attempting to reflect the real-world situation where such differences are to be found. By the time of the 2007 drought, the sheep numbers on this property had dropped to 8000 and cattle to 150. The number of bales of wool produced went from 360 to 240. The wool clip loss is therefore 120 bales at an average price of $1700, minus the expenditure saved by running and shearing fewer sheep (i.e. $204 000 minus $90 000, which results in a $114 000 loss). If most of the 3000 sheep were sold in that year, there would be the income from the sale in that year. However, in future years the impact of the destocking would be significant. The money from the stock sale would increase farm income only in the year of sale. Destocking is what economists call disinvestment. The value of the farm including its stock will drop in the next year and subsequent years. Again we have neglected the decreased cattle numbers and, most importantly, the ongoing loss in future years. However, we imagine you are getting an idea of the losses involved and the dynamic nature of the losses.

And yet another hypothetical farm, this one of 1000 ha, would normally run 4500 sheep, with approximately an equal numbers of ewes and wethers/dry ewes. No cattle are run. The stocking rate equals 4.5 sheep/ha. A total of 200 bales of fine to medium wool are produced, which means on average 22 fleeces per bale. This is a significant difference in the fleece weight compared with our other farms, yet such differences are to be expected – some sheep are larger and carry more wool. One of the values of using hypothetical examples is the ability to show such significant differences based on real-world data.

As a consequence of the 2007 drought, a reduction of the flock by 450 results in 4050 sheep. This is an insignificant loss compared to some; very wide differences were to be found in the catchment. For this farm the stocking rate falls to 4 sheep/ha. The number of bales produced falls to 170, now with 24 fleeces per bale. The average fleece is marginally lighter due to the drought. Viewed in these terms the drought caused the loss of production of 30 bales of wool. There had been a smaller reduction in the previous year (2005–06), while the year 2004–05 had been a normal year. The loss of 30 bales (in this case, at $1500 per bale) results in a figure of $45 000. There are 450 fewer sheep to shear, which reduces the farm costs by $13 500. This gives a net loss of $31 500 in 2007. Matters were worse for this (hypothetical) farm than the others we have described as the drought had a serious impact in the previous year, which was not the case for the others. This impact has not been calculated.

The cumulative costs of long-running droughts lead to a greater loss than the one-year loss estimated in our examples. The total value of lost production has to be counted over all the years when losses occur.

The loss of wool production on the farm we have just discussed occurred, notwithstanding an increase of area sown for crops. The crop sowing was an additional cost that would have to be included – even though the crop failed. The crop yield declined a little in 2004–05, and then declined in 2005–06 to a disastrous level; next came the extreme conditions of 2006–07. In this situation there was not only the loss of wool but the additional cost of 20 ha of wasted crop planting. While a hypothetical example, there were farmers we interviewed who had experienced somewhat similar, but not identical, expenditures.

We have been discussing farms running both sheep and cattle, and the sheep comprise ewes, wethers and some lambs. The calculations have been made as simple as possible by estimating nothing more than the loss of wool revenue. A more complete analysis is required for the mixed cattle and sheep farms. Cattle require more land per beast than sheep, just as breeding ewes require more food than wethers. There is a standard measure by which to convert those different animals and their nutritional maintenance requirements to a equivalent measure (a measure of equivalence). It is the 'dry sheep equivalent' (DSE), which is the maintenance requirement of a 48 kg wether for one year. As sheep and cattle have different grazing patterns, and grazing is an ever-changing activity, more precise measures than the DSE are used to determine the energy supplied to animals by a variety of feeds and the expenditure of energy by the animals. However, DSE is the conventional rule-of-thumb measure. For simplicity in calculation and ease of explanation, we rely on DSE in our research.

The impacts of the drought: in summary

During the drought, farmers found that many of their water-holes had completely dried up. There was no water for those who irrigated, and no water to cart to outlying paddocks. As a general response to the drought, farmers attempted to keep breeding ewes and breeding cows while sacrificing other animals. With some exceptions, the numbers of breeders were kept constant. However, the number of wethers and dry ewes were reduced to under half the normal number. The number of steers and stores were similarly reduced. The number of lambs weaned fell to 89.9 per cent of the normal percentage.

A very significant number of farms (over 85 per cent) were feeding stock, with some commencing this in late 2006 as farmers' own food reserves were soon depleted. Crops failed. About one-quarter of pastures will need to be resown at the end of the drought as they have become degraded. This is the dynamic of a severe drought. Grain was fed to stock at a significant cost. Other types of farming suffered. Potato yields were reduced for those who relied on this form of agriculture.

The most marked costs of the drought have been the loss of productivity in wool growing. This is why we have used wool damage costs to illustrate our findings. Most wool producers experienced a reduction to 75 to 85 per cent of the normal clip. The forgone income was substantial – ironically it was greater in 2008 than the same loss of productivity would have been only two or three years earlier when wool prices were considerably lower. Wool had become more valuable in the market; hence losses more significant. Here was a double whammy for those who thought about it. This result helps us make a very basic point. The value of water to a farmer (and to others) is derived from the value of what it is used to produce, and if the value of the latter changes due to exogenous factors (displacement by a new product) the value of water in this use changes. The variability in market prices – of so many goods and services – is the major impediment of developing economic–environmental links. This is of such importance that it deserves pride of place in the results from this research. We started out to make the link between the environment and economic values, only to come to the conclusion that one variable is far less open to changes except over long periods (the environment), while economic values can be highly variable over short periods of time, due in large to exogenous factors. This means that we can end up with a range of snapshots of the relationship, each depicting the exogenously determined value of the economic good. It is only if we can remove the exogenous influences on economic values that we can establish constant (long-term) environment–economic relationships.

It is opportune to remind the reader of the near impossibility of coming to *one* number for the value of water to a farmer – that value changes as the value of the farm products change. In the case of wool, the increase in its value was significant and the value of water increased. A significant cost to most of the farmers was in planting feed crops for the animals.

From the loss of wool production we get the initial, minimal impact of a reduction in water availability – in other words the value of the water that was *not* received is the net value of the forgone wool income – a loss in the order of 20 per cent based on the normal year. A property producing 100 bales in a normal year produced 80 bales in 2006–07. The gross value of this lost production was $30 000 to $40 000, while the net value was approximately half of these amounts. The bales *not* produced would have been more valuable due to the price increases. Here was a double whammy for Little Swanport farmers. Put in other words, assume that the same amount of water had been available to farmers in 2006–07 as was available in a normal year; the value of water is given by the net loss due to the difference in water availability. *Here we have the value of water in economic terms to farmers.*

On the basis of the additional data we gathered in the drought year 2007 we derived estimates of the net economic costs of the reduced rainfall in the years 2005–06 and 2006–07. The estimates are based on the impact on the total farming community in the catchment, recognising that rainfall – and the response to the lack of it – varied significantly across the catchment. Our estimates do not include the future costs, something that we have drawn attention to but not calculated. A comprehensive analysis of the damage costs (resulting from less water) would estimate the losses year-in year-out until normal economic earnings returned, and when we were writing this that point in time was not obvious. An extremely important feature of the damage caused by droughts is that it does not happen immediately (that is, in the year of the drought), and the damage is cumulative in as much as one impact, for example poor lambing, results in decreased wool clips and meat sales well into the future. This effect can be mitigated by immediately restocking, which comes at a cost.

The costs we have estimated are based on three different measures used extensively in the environmental economics literature: replacement costs, preventative costs, and lost productivity costs (damage costs). For the year 2005–06, the total across-the-catchment loss is estimated at $1 553 000. For 2006–07, the estimate is $3 355 000. We can compare these economic losses to rainfall data. These losses to farm income resulted in reduced purchase of goods and services in the local towns. We have shown earlier the extent of economic reliance of these towns on the catchment. In the drought years, these towns lost income from the catchment farming community; however national population growth and tourism to the area compensated to varying degrees – much depending on the nature of the business. A more comprehensive analysis would delve deeper into the circumstances in the towns.

Estimating the value of water when it does not rain

We now have considerable data from the catchment farmers on how they have attempted to deal with a very serious drought. The data we have shows the monetary loss starting from a position of normal rainfall and declining, in our case dramatically.[6] Sometime in the future we could undertake the much more simple exercise of recording the increase in profits that result from the added quantity of water, as rain, that returns the situation in the catchment to normal.

We are going to use rainfall, which is conventionally measured in millimetres, rather than water captured, which is conventionally measured in litres, to obtain an estimate of its value to the farmers. The use of millimetres is appropriate in our case because what we have in the catchment is rain-fed agriculture with only limited irrigation of crops. While the larger farms are a combination of improved pastures and run country, and the improved pastures are obviously fertilised and seeded, they rely on rain for grass growth. If (as was proposed) there existed a very large water storage, or even a number of significant ones, that extracted water from the Little Swanport River, we would use the reduction of water due to these impoundments as an independent variable in our analysis. The fact that we have only small farm water-holes (even though here are approximately 1200 of them) gives us no scope to use changes in volume of water impounded. The Tasmanian Government has no record of quantities of water used by farmers.

The non-terrestrial catchment farmers, the oyster growers in the estuary, rely on water in a different form, in a flow across their oyster racks. Of course, the freshwater flow in the estuary is a function of rainfall throughout the catchment, plus any groundwater entering the main river and its tributaries before they reach the estuary, minus any water extractions that reduce the downstream flow. We have no data on the groundwater in the river basin. Whether or not the Little Swanport is a 'losing' or a 'gaining' river would be an important consideration in comparing upstream to downstream water flows.

In making our calculations we use the Bureau of Meteorology records for years when rainfall change was most evident – 2005, 2006, 2007 – plus the moving average. The latter measure is indicative of what we have been calling a normal year. The rainfall records are taken at three recording stations just outside the catchment. The first is at Ross in the north-west, the second at Swansea at the north-east, the third at Orford at the south. As is obvious in Table 17.2, the average rainfall varies across the catchment with the south being considerably wetter than the inland top of the catchment. However all of the catchment is

Table 17.2. Rainfall: selected sites, 2005 to 2007

	Rainfall (mm)			
Site	**2005**	**2006**	**2007**	**Average as of 2005**
Ross	630	357	353	503
Orford	727	408	405	679
Swansea	621	335	353	597

dry – and hence is good Saxon merino country in normal conditions. The fact that Australia's mainland merino sheep are grazed west of the Great Dividing Range is recognition that this breed of sheep is prone to foot rot in high rainfall environments.

The data we obtained in our initial survey of farmers, which covered two non-drought years and the first year of the drought, showed far less damage costs than did our follow-up survey, which captured the consequences of the second year of the drought. In the first drought year there were some replacement/preventative expenditures undertaken. Farmers were for purchasing more stock feed than normal, growing more feed on the farm than normal, plus some other similar types of costs. The total cost (which means a loss) for the catchment farmers was $1 553 000. This was for the year 2005–06.

Far greater damage costs – which commenced with the drought conditions in 2006 – showed up in the year 2006–07. In that year the replacement/preventative expenditures increased as farmers became more desperate in their attempt to reduce the extent of the damage and, more importantly, there were damage costs due to a loss in wool production, in fat lamb sales, and in beef production. We did not obtain sufficient data on the loss of cash-crop income (mainly potatoes) to document the loss on this aspect of farming; however, our available data covers all grazing properties plus the sheep/cattle component of the mixed farms. It is an underestimate – not only for the reason that cash crops are excluded but for the extremely important reason that we have not considered anything other than one year. We have discussed this limitation already.

Before presenting the data for 2006–07, it is very important to emphasise (once again) the fact that wool and meat prices increased substantially in the latter part of the decade. Fine and extra-fine wool (the 'bread and butter' of the catchment) reached price levels not seen for a considerable time. These price increases cushioned – and quite significantly – the immediate adverse impacts of the drought; however, they increased the forgone income of drought-affected farmers. Farmers were obtaining higher prices for their meat and wool, yet their potential income was reduced – the opportunity cost of forgone production. For an example, a reduction in the wool clip where the selling prices is 1700c/kg is a much greater loss than if it was to sell for 850c/kg.

There is not one consistent story in the catchment. Particular farms, which by the vagaries of nature have obtained more (of the scarce) rainfall than others, endured good economic times; however, increased fuel and feed costs had a dampening effect even on these farms.

We can measure the *value of rainfall* as damage costs, using a range of selling prices for wool and meat, and this would be a sensible approach taking the analysis into future years. As a starting point it is illustrative to present the damage costs as they occurred in 2006–07 (where good prices prevailed).

Our data indicates that the loss of productivity for wool and meat combined (sheep and cattle) was in the order of $1 575 000 in 2006–07. The total of replacement/preventative expenditures during this year was approximately $1 780 000. The total damage cost was the sum of these two, $3 355 000. Given that rainfall over the catchment was about 50 to 60 per cent of the normal year – and this was the case for two years running – the reduction in income to the catchment was very high and approximately one-third of its normal state.

This should not be taken to infer that there is a linear correlation between rain and income. More importantly, it should not be assumed that even worse drought conditions than those experienced in 2006–07 and 2007–08 would result in a further linear decline in income. At some low level of rainfall the whole economy of the catchment tips to an unknown future. Where wool and meat prices remain high at their recent maximum the loss is greater than it is with a lower price, reaching in the order of $4 million per year.

What we have done is show the correlation between rainfall – as it falls over the entire catchment – and economic returns from the terrestrial farming in the whole catchment. That is an important piece of information. The less rain the less grass, the less wool and meat for the market. The value of crops like potatoes can be correlated with rainfall. We have not estimated this. We have also said very little about rainfall and the oyster farmers who are making a living at the bottom of the catchment.

Oyster farming is a minor, but profitable, fishery in Tasmania. The value of oysters produced each year in the state is approximately $20 million. Of that, the Little Swanport estuary oyster production is significant, not only in the production and sale of mature oysters but in the production of immature oysters for sale to growers elsewhere in southern Australia. Oysters in the range of 20–30 mm are in high demand to be on-grown. There is yet a further step back in the chain. The firm Shellfish Culture, located in the Little Swanport estuary, produces a substantial proportion of the Pacific oyster aquaculture industry requirements for immature oysters for on-growing in Tasmania and in South Australia. Additionally, Marine Culture, one of the oyster farming companies operating in the estuary, regularly transfers immature oysters from its farm in Little Swanport to other southern Tasmanian farms for grow-out to market size. At 'farm gate' prices, the Little Swanport estuary is capable of producing up to $7 million worth of oysters per year,[7] with seed oysters contributing more than half. Our modelling research has indicated that there could be a loss of production of mature oysters of 12 per cent if fresh water availability in the estuary were reduced by 40 per cent. This would involve a loss of $500 000 per year at the prices that prevailed when we gathered our data from the oyster farmers. Either a severe drought or the construction of a large impoundment upstream of the estuary could – while the drought persisted and/or until the dam water reached its maximum before any water was released – cause a reduction of water in the estuary. What the economic value of water in a large impoundment would be can only be answered by reference to the use that water was put to. A large dam would not be of benefit to the grazing industry in the catchment as it requires a good spread of rainfall across the catchment.

18

Conclusions

Christine Crawford and Tor Hundloe

We have come to understand the human–environment dynamics of a small Tasmanian catchment. Our focus throughout the extended period we spent in the field (and sitting in front of computer screens) was to explore the relationships between the use of water by farmers and the needs of oyster growers. The relationship between upstream catchment activities and the quality and quantity of water downstream (in estuaries and off-shore lagoons) is the fundamental issue in catchment management. It necessitates an analysis of the value of water in different, and often competing, uses. On the face of it, that might seem a simple exercise. We show that it is complex, particularly once the researcher widens his/her horizons and looks to the people and industries who are not part of the catchment community but who rely on the catchment for their livelihood.

An important difference in the value of water to agriculture and aquaculture production that became apparent during our study was that agriculture largely relies on the quantity of water available, with essential nutrients provided in the soil or artificially from agricultural fertilisers. Aquaculture production, however, is reliant on the quality as well as the quantity of water flowing into the estuary. The freshwater contains nutrients, mainly nitrogen and phosphorus, to support the production of the phytoplankton food of the oysters, while the quantity of water delivered influences the amount of time that the nutrients remain in the estuary, and hence the time available for biological uptake. Silicates delivered in the freshwater are also essential for the production of diatoms. This issue of quantity versus quality makes comparisons between the different users even more complex.

For both the agricultural production in the catchment and the oyster production in the estuary, the drought conditions experienced during the three years of this project were initially problematic because occasional flooding was considered to be an important component of the annual water cycle, and we wanted to compare production between normal flow and flood conditions and to assess the importance of increased freshwater flow to the estuary. In the end, however, we used the drought conditions to provide a comparison of the value of water to agricultural and aquaculture production when water was plentiful with the value when it was severely limited during drought years. This provides estimates of the economic value of water at two (extreme) points on a continuum. It is not necessarily the case that the relationship is a linear one; however for policy-making purposes it could be assumed as such, particularly if an error band around the estimates is allowed for.

Across the catchment the loss of income to the catchment from wool production, fat lamb sales and beef production when rainfall was approximately 50–60 per cent of a normal year was estimated to be \$3.36 million, or approximately one-third of normal income from wool, fat lambs and beef. When the prices of these farm outputs increase, as they did near the end of our survey period, the dollar loss is greater; conversely, if farm product prices drop, the loss from droughts is smaller. (Cash crops were not included as there were insufficient data.) These values are determined from the sum of preventative expenditure, replacement costs and loss of production incurred due to the drought. In the estuary, nitrogen budgeting indicated that the increase in oyster harvest across two wet years (2004 and 2005) was estimated at approximately 43 kg N or a 12 per cent increase relative to the two drought years of 2006 and 2007. This equates to a loss of approximately \$500 000 in a severe drought year. Both terrestrial farming and oyster growing lose in severe drought years. If we take as the normal yardstick the pre-drought situation, we are able to say that a reduction in water in the catchment of about 45 per cent results in a loss of income to all major primary producers of approximately \$4 million per annum. If the relationship was linear, a decrease in water availability of about 11 per cent would cost primary producers \$1 million annually.

The loss in production in the estuary during the drought that prevailed over much of the study period was largely due to a lowering of the growth rate of the oysters; mortality was not observed to increase. Sufficient food was provided from oceanic waters and produced in the estuary to maintain the oysters, but they were slower growing. During the drought, with minimal freshwater flow into the estuary an influx of nutrients during winter indicated that nutrient-rich Southern Ocean waters penetrated up the east coast of Tasmania and into the estuary. Thus, as a consequence of the drought, the oysters took longer to reach market size and condition. On land, however, many farmers were forced to destock. The condition of grazing land was degraded in some areas as farmers struggled to maintain their herd. Crops either failed or produced less than normal, and were not sown due to lack of water storage. Thus, the recovery time after the drought is likely to be greater on agricultural farms, taking several years to improve the condition of the grazing land and to restock with cattle and sheep, whereas in the estuary the recovery time is in the order of months. Recovery time also depends on the stocking density before the drought and whether the farmers were stocked to full capacity for good growing conditions or whether they maintained a lower stocking level that would provide a buffer during drought conditions. In the estuary, where food for the oysters is produced throughout the estuary and transported across the farms by the tide and wind-driven water currents, the oyster farmers have reduced their stocking density and area available for farming so that they can produce a marketable product for most of the year under a variety of conditions. This is, clearly, a sound business strategy.

In relation to environmental flows to the estuary, it is important to note that maintaining the low flows is most important. The transport model that was developed for freshwater flows through the estuary predicted a nonlinear response to increasing river flow; flushing time (the amount of time that the freshwater and the nutrients it contains remain in the estuary) decreases quickly as river flow increases, but is relatively stable at high river flows. Ecosystem model simulations at different levels of base flows, based on this transport model, thus predicted that phytoplankton biomass, and consequently oyster growth, initially increases rapidly with base flow before the rate of increase slows to a steadier rate at higher flows. Therefore, there are greater benefits to the estuary per ML of river flow at low flow than at high flows. At low river flows, phytoplankton have more time to take up the additional nutrient inputs from the river because the time to pass through the estuary is longer. In contrast, at higher flows there is less time for biological uptake as the flushing time is shorter, and so the benefits are smaller per ML of river flow. For example, as base flows increase from 1 to 40 ML/day, oyster growth is predicted to increase by 11 per cent, but another 11 per cent increase in growth occurs only when the base flow reaches 200 ML/day. The results of this study therefore support the cease-to-take requirements for low flows in the Water Management Plan for the catchment. However, the modelling predicted that the greatest benefits from river flow are achieved over the summer months because higher water temperatures significantly enhance the growth rates of phytoplankton and oysters.

The increased allocation of water for stock, domestic and irrigation purposes in the Water Management Plan (2006) from 3882 to 6084 ML/year was shown by modelling to be unlikely to have a significant impact on the estuary for average and dry years. Modelling conducted using very dry flow conditions, as experienced in 2007, however, predicted that water harvesting during summer to the increased full allocation would lead to a decline in estuarine nutrient and phytoplankton concentrations and hence oyster growth responses. Taken across the whole year the changes due the increased allocation would be relatively small, and would fall within the limits of uncertainty inherent in the model simulations. Nevertheless, these results do imply that harvesting water during a very dry year is more likely to affect the estuary, especially during the summer months.

Although this research has centred on the Little Swanport catchment, the techniques developed are of relevance to the assessment of catchments anywhere. This is particularly the case with the water budgeting and economic and social analyses. The biogeochemical model can be applied in other estuaries where there are sufficient local data, particularly in terms of hydrodynamics. The nutrient budget process can also be used in other estuaries with relevant local nutrient data available. The water evaluation framework developed for the catchment provides a generic template for catchments to assess the value of water to different users across a catchment. To guide others who wish to understand water catchments in a holistic sense, we point to the data requirements, survey methods and types of analyses required, along with the potential difficulties that are likely to arise.

The results clearly show that the profitability of both agriculture on land and aquaculture in the estuary are affected by changing freshwater flows. With changing climate regimes and predictions for drier conditions in some regions of Tasmania, as well as more extreme events, catchment communities will need to work closely together to protect and share scarce water resources. Having a better understanding of the quality and quantity of water across the catchment and accounting for water use will be essential to this process.

In summary

We have been able to calculate the economic damage costs to both terrestrial farming and aquaculture in the Little Swanport catchment that result from the recent extremely severe drought. Less water falling as rain resulted in a loss of production (of wool, meat and crops), and decreased flows in the estuary resulted in the loss of output in oysters. As terrestrial farming is a much larger industry, the magnitude of the loss to this sector was higher than that to the oyster industry. With rainfall at between 50 to 60 per cent of an average year, the costs to terrestrial farmers were found to be in the order of one-third of net revenue. The costs to the oyster farmers of lost production from reduced flows were estimated (from a modelling study) to be in the order of one-sixth to one-twelfth. The former estimate is based on the sale of mature oysters while the latter is based on the sale of mature oysters plus seed oysters. Put in other words, for every dollar of output from terrestrial farms the drought resulted in a loss of 33½ cents, while for every dollar of output of mature oysters the loss was 16⅔ cents.

The above estimates represent two points on a curve – a state of normal rainfall and a reduction of between 40 and 50 per cent. We would like to join these two points and illustrate the damage costs in small (marginal) steps. We hesitate to do that as there are thresholds that suggest that this is not a smooth relationship.

We must restate the most serious caveats we have. First, the damage cost for wool and meat production varies according to the market price of these products. At much lower prices for wool, fat lamb and beef, as have prevailed in the past, the loss to terrestrial farmers is considerably less. What is true for these farmers is true for oyster growers. Second, the losses to the oyster industry are based on a modelled simulation of oyster productivity under different river flow conditions, and do not include any operational costs.

We have done more than delineate a draft water account for the catchment and link rainfall (water availability) to farming and aquaculture. We have identified the extent to which the adjacent towns rely on the catchment economy, an important relationship. We have explored the social capital of the catchment communities and found it to be high, and we have identified and measured some of the non-market values that attract people to the catchment – and provide an incentive for them to remain there.

Notes

Chapter 1

1. Its location (assuming there was such a place as identified in the Old Testament) is contested. Conventional wisdom suggests that it was in present-day Iraq, maybe at the head of the Persian Gulf where the Tigris, Euphrates and two other rivers meet.
2. The invasion of Iraq by the United States and its allies took place in 2003.

Chapter 2

1. The World Bank has changed the terminology it uses to describe the various types of capital. The bank categorises the forms of capital as natural capital, produced capital (our human-made capital) and intangible capital (social capital plus human capital). We shall continue to use the previous definition because it provides greater clarity.

Chapter 3

1. Inflation is accounted for by changing nominal prices to real prices, using a simple formula available from the Australian Bureau of Statistics or its equivalent in other countries.

Chapter 4

1. The few remaining absolute monarchies – in the Middle East and Brunei – are the extreme end of a continuum. The pre-eminent sheik or sultan owns everything.
2. Damage costs are not confined to water (our interest here) but to all circumstances where production, profits and income are decreased as a result of an event outside the control of the input owners. For example, upstream soil erosion on farms that leads to increased sedimentation downstream and reduced incomes for fishers and others is an example of a damage cost.
3. For very small river basins (which could be sold as a unit) we could use market prices to value the whole catchment.

Chapter 5

1. The economic rule is to keep adding an input while the value of its additional product remains greater than the cost of the input.

Chapter 6

1. Large cities are classified as having a population greater than 100 000 people.
2. Alexander the Great ruled from 336 to 323 BC. He died aged 33 years.
3. That is, dams more than 15 m high.
4. There are 34 rivers in the world that are larger.

Chapter 7

1. For example, the REALM model is a major water cycle model and is often used as a primary tool for water balance modelling that predicts impacts of water use scenarios on stream flow.
2. The most relevant studies of the catchment include the hydrological assessment in reports by the DPIWE (2003) and DPIW (2007a; 2007b) plus a water balance assessment of the catchment covering both natural supply-oriented climate and hydrological aspects, as well as socioeconomic and institutional dimensions such as water entitlement, use, and farm dam water use and flow impacts under a variety of policy or regulatory scenarios (Sinclair Knight Merz 2004). There are also a number of other useful reports containing water-related and detailed land use and socioeconomic data utilised in the preparation and review of the Little Swanport Catchment Water Management Plan (for example, Little Swanport Catchment Committee 2006; Resource Planning and Development Commission 2006).
3. Examples include the Common Chart of Water Accounts, the Bureau of Meteorology data system, South-East Queensland's *WaterHub*, earlier land and water audit-based approaches (such as the NWC's (2008) Australian Water Resources 2005), and Australian Bureau of Statistics' water accounts (ABS 2006a).
4. The data collection process for Australia's National Water Accounts is likely to be administered under the Australian Government's Bureau of Meteorology (BOM). The BOM is already preparing its water information system as required under Section 126 of the *Water Act 2007*.

Chapter 8

1. The small team (described later) has spent the best part of three years researching the catchment – and still there are major matters deserving more investigation. The fact that teams of scientists much larger than ours have attempted, over many decades, to make ecological and economic sense of the Murray–Darling basin, with only limited success, puts our effort into perspective.

Chapter 9

1. Waterways Monitoring Reports, available at <www.dpiw.tas.gov.au/inter.nsf/WebPages/JMUY-6BV8GJ?open>

Chapter 10

1. In comparison to similar systems in New Zealand; McIntosh *et al.* (2005).

Chapter 11

1. The detailed description of the ecosystem model or the methodology and analytical procedures of field observations is provided in a report to one of the funding agencies that has supported this project, the Fisheries Research and Development Corporation (Crawford *et al.* 2010).
2. These estimates are based on the field observations and a full description of the calculations can be found in Crawford *et al.* (2010)
3. See Crawford *et al.* (2010) for a detailed description of the calculations used to estimate river loads.
4. See Mann and Lazier (2006) for a detailed explanation of estuarine circulation.
5. See Crawford *et al.* (2010) for a detailed description of model calibration and validation.

Chapter 12

1. Sixty-five, according to Nyman (1990, p. 35)
2. Feeding took place twice a day during winter. Chopped-up turnips and swedes were mashed with chaff and taken to a small, sheltered paddock near to the milking bales where the cows stayed overnight.
3. The energy source for the machinery was locally sourced wood. Water for the boiler was also obtained locally, yet again an example of nature providing ecosystem products without charge.
4. Due to the faults all historical records fall subject to, particularly the fact that original deeds of grant no longer exist.

Chapter 13

1. We should note that the census population number is considerably less than the 700 people we find referred to in various official documents. The population in the catchment (this is the geographical catchment or the river basin, not the larger area used by Tasmanian authorities) has been estimated using 'mesh blocks': small land-units (smaller than census districts) used by the Australian Bureau of Statistics. These are the smallest units used for statistical purposes in Australia. Only population data can be obtained at this level as no other information is gathered. To the authors' knowledge this is the first-ever reporting of mesh block data for Tasmania, and hence the first accurate population count of the Little Swanport catchment.

Chapter 14

1. Fishing offshore produces a greater range of species, although flathead dominate. On average the number of days involved is 20, which is split approximately between regular fishers (from 20 to 60 days per year) to occasional fishers (1 to 15 days per year). Twice as many flathead are caught offshore than any other species. Other species include couta, cod, salmon, squid, crayfish and trumpeter, and the harvest of abalone.
2. See Hundloe (2002) for a detailed discussion and an example.
3. See Hundloe (2002).

Chapter 15

1. For more detail on the agricultural census see <www.abs.gov.au/ausstats/abs@.nsf/dossbytitle/AD7C6DD1D14FB809CA256BD000272737?OpenDocument>. Note that the ABS states that 'for the 2005–06 Agricultural Census, the ABS undertook a special project, with additional funding provided by the National Water Commission. This project involved coding all farm businesses to a relatively precise location, allowing for the production of estimates by customised boundaries. It is not certain whether or not this work will be continued for future Censuses' (*pers. comm.*).
2. A less reliable estimate based on the ABS Small Area Agricultural Statistics (Cat. No. 7125.) suggests that there could be 250 000 sheep in the catchment.
3. I thank Dr Julian Morison for advice on the magnitude of Tasmanian multipliers.
4. We do know that some owners (in our case farmers) are better business people than others and, hence, operating expenses and profits will change with a change in ownership.
5. It is marginally higher. We have rounded the numbers to the nearest $5.
6. As determined in the census of August 2006.
7. Teenagers spend two-thirds the amount adults spend on the items under discussion.
8. Converted to full-time residents. The vast majority of non-permanent residents are land or land-and-home owners in the lower catchment villages of Pontypool and Saltworks Road.

Chapter 16

1. It is appropriate to reiterate that a major although small-scale irrigation source of water in the catchment is Hobbs Lagoon. This water is sourced from outside the Little Swanport catchment (from the Prosser River catchment) and its related flows have been ignored for practical purposes in the water accounts although, technically, they should be included as system imports and exports.
2. These quantities have been gleaned from a variety of sources.
3. Human consumption of water (calculated as around 5 GL per annum) is not identified within this general breakdown but would be partly accounted for in the total evaporation and transpiration levels.
4. For example, around 50 per cent of the catchment surface vegetation and cover (and soil and runoff conditions) has been modified for grazing and other economic purposes and significant remaining sections have been impacted by humans at some time in the past.
5. They have been based on generalised parameters from water budgets for the Border Rivers–Gwyder region in New South Wales (Border Rivers–Gwydir Catchment Authority 2008) and adjusted in accordance with estimates of 10 per cent runoff revised up to the relatively reliable 22 per cent runoff derived for this catchment (113 GL of 500 GL). This annual outflow measure is based on results presented in DPIWE (2003). As noted previously it is derived from data for 1971 to 1990 and is unlikely to represent low-flow periods of more recent years. Some more detail on estimates for overall levels (and most account items) is provided in the Notes column of Table 16.1.
6. Sources for these estimates are provided in Table 16.1.
7. While 113 GL is noted by DPIWE (2003) as a minimum figure, it must be remembered that this estimate is based on flow records between 1971 and 1990 and rainfall in the catchment decreased substantially in the 15 years following this period.

8. However, it is unlikely that 100 per cent of the water extracted by households would have contributed to surface waterway flows. It is probably closer to the 25 per cent of rainfall that enters watercourses across the region. This adjustment has not been included to keep the analysis simple (and in view of the low level of flows involved).
9. Note that water use from dams is not based on farm dam size (which obviously has no direct relation to annual consumption) but has been calculated by Sinclair Knight Merz (2004) via survey-based experience with usage rates based on water use type. The actual water applied to irrigation and stock and domestic water has been estimated as total surface water flows into dams (4.92 GL) minus dam evaporation (960 ML) and groundwater seepage (492 ML). Evapotranspiration and groundwater seepage from applied water is calculated by a ratio to this quantity.
10. While we are using sheep to reflect the demand for water by domesticated animals, we could just as easily have used cattle, or any mix of cattle and sheep.
11. In other studies of catchments we would be unlikely to be as concerned with severe drought conditions as this one. We would be willing to accept long-run averages. The recent drought in the Southern Midlands of Tasmania was one of the most severe on record.
12. There are returns to the catchment from irrigation from Hobbs Lagoon (with its water extracted from outside the catchment). However, these returns have not been included in the analysis so existing net extraction levels may be lower (possibly by 500–1000 ML). There is some debate as to how much water is actually transferred from Hobbs Lagoon to the Little Swanport catchment.

Chapter 17

1. Very occasionally a disaster caused by a combination of an extreme weather event and human failure, such as grounding an oil tanker in a cyclone, has a temporary effect on local industries. Likewise, the very occasional extreme natural event, such as the 2006 tsunami in South-East Asia, can have a disastrous impact on the tourism industry. Such events are rare.
2. One of the editors of this book was a Visiting Scholar at the East-West Center in the 1980s when much of this work was done, and remains an alumnus.
3. Where actual market information is not available (revealed preferences), economists have to rely on 'stated' preferences – asking respondents what they will/might do – or derive estimates from the expenditure of respondents in other surrogate markets.
4. While we are using the term 'the 2007 drought' to describe a drought event, it should be recognized that the previous year (2006) was also a drought year and reductions in stock numbers (particularly beef cattle) had commenced in 2006.
5. This would affect the farmer's credit worthiness in the eyes of the banks, putting additional pressure on the farmer.
6. In fact the year 2005 was one of above-normal rainfall in the catchment and this we have taken into account.
7. The major flow-on economic impact of oyster growing in the catchment results from the mature oyster market.

References

ANZECC (2000) 'Australian and New Zealand Guidelines for fresh and marine water quality'. Australian and New Zealand Environment and Conservation Council, National Water Quality Management Strategy Paper No. 4. *Australian Journal of Marine and Freshwater Research* **38**, 261–288.

Australian Bureau of Statistics (2006a) 'Water Account Australia 2004–5 Cat. No. 4610.0'. AGPS, Canberra.

Australian Bureau of Statistics (2006b) 'Water Use on Australian Farms 2004-5'. Cat. No. 4618.0. Comm. ABS, Canberra.

Australian Bureau of Statistics (2007) 'An experimental monetary water account for Australia'. Cat. No 4610.0.55. ABS, Canberra.

Ayres R (1978) *Resources, Environment, and Economics: Applications of the Materials/Energy Balance Principle*. John Wiley & Sons, New York.

Bain D (2008) 'Climate change and water: how water accounts can help our understanding'. UNSD Conference on Climate Change and Official Statistics. Oslo, April 2008. Paper for Session 3. Centre of Environment and Energy Statistics, Australian Bureau of Statistics.

Beck MW, Brumbaugh RD, Airoldi L, Carranza A, Coen LD, Crawford C *et al.* (2011) Oyster reefs at risk and recommendations for conservation, restoration, and management. *Bioscience* **61(2)**, 107–116.

Border Rivers–Gwydir Catchment Authority (2008) 'Calculating your own water budget'. New South Wales Government, <http://brg.cma.nsw.gov.au/index.php?page=calculating-your-own-water-budget>.

Boyce J (2006) Canine revolution: the social and environmental impact of the introduction of the dog to Tasmania. *Environmental History* **11**, 102–129.

Boynton WR, Kemp WM and Keefe CW (1982) A comparitive analysis of nutrients and other factors influencing estuarine phytoplankton production. In: *Estuarine Comparisons*. (Ed. VS Kennedy) pp. 69–90. Academic Press, New York.

Brown MR and McCausland MA (1999) 'Increased production of juvenile Pacific oysters (*Crassostrea gigas*) through supplementary feeding'. CSIRO Marine Research, Final Report to FRDC, Project 94/83, Hobart.

Bureau of Rural Sciences (BRS) (2008) July 2008 Drainage Division Summary. Tasmania. CSIRO. Bureau of Meteorology. <http://adl.brs.gov.au/water2010/pdf/monthly_reports/awap_1003_report.pdf>

Clementson LA, Harris GP, Griffiths FB and Rimmer DW (1989) Seasonal and inter-annual variability in chemical and biological parameters in Storm Bay, Tasmania. 1. Physics, chemistry and the biomass of components of the food-chain. *Australian Journal of Marine and Freshwater Research* **40**, 25–38.

COAG (Council of Australian Governments) (2004) Intergovernmental Agreement on a National Water Initiative. <www.coag.gov.au/meetings/250604/index.htm#water_initiative>

Costanza R, d'Arge R, de Groot R, Farberk S, Grasso M, Hannon B. (1997) The value of the world's ecosystem services and natural capital. *Nature* **387**, 253–260.

Crawford C, Hundloe T, Ross J, and Fulton B (2010) 'Water use across a catchment and effects on estuarine health and productivity'. Final Report to the FRDC, Project 2005/072. Tasmanian Aquaculture and Fisheries Institute, University of Tasmania, Hobart.

Crawford CM (2003) Qualitative risk assessment of the effects of shellfish farming on the environment in Tasmania, Australia. *Ocean and Coastal Management* **46**, 47–58.

Crawford CM, Macleod CKA and Mitchell IM (2003) Effects of shellfish farming on the benthic environment. *Aquaculture* **224**, 117–140.

Crawford C and Mitchell I (1999) 'Physical and chemical parameters of several oyster growing areas in Tasmania'. Tasmanian Aquaculture and Fisheries Institute, Marine Research Laboratories, Technical Report Series, Number 4, Hobart.

Crawford C, Mitchell I and Brown A (1996) 'Predictive modelling of carrying capacities of oyster (*Crassostrea gigas*) farming areas in Tasmania'. Marine Research Laboratories, Final Report to the Fisheries Research & Development Corporation (FRDC), DPIF, Hobart.

Crawford C, Neira FJ and White C (2006) 'Environmental flows in the Little Swanport estuary'. Tasmanian Aquaculture and Fisheries Institute, NHT Final Report 46932, Hobart.

Crosby MP and Gale LD (1990) A review and evaluation of bivalve condition index methodologies with suggested standard method. *Journal of Shellfish Research* **9**, 233–237.

Dare D (1989) *A History of Woodsdale*. History Society of Leverdale and Woodsdale, Woodsdale.

Dare D and Bellette J (n.d.) So *Far From Home: Early Settlers of Woodsdale*. History Society of Levendale and Woodsdale, Woodsdale.

DEWHA (n.d.) Register of the National Estate Database Place Report: Jacks Island Area, Database No. 011743, Canberra.

DHHS (2010) Little Swanport triennial data review 2010. Tasmanian Shellfish Quality Assurance Program, Department of Health and Human Services Tasmania. <www.dhhs.tas.gov.au/peh/tsqap/publications/sanitary_surveys2DHHS 2010>.

Dinamani P (1987) Gametogenic patterns in populations of Pacific oyster, *Crassostrea gigas*, in Northland, New Zealand. *Aquaculture* **64**, 65–76.

Dixon JA and Hufschmidt MM (1986) *Economic Valuation Techniques for the Environment*. Johns Hopkins University Press, Baltimore.

DPIW (2006a) 'Little Swanport catchment water management plan'. Water Policy and Planning Branch, Department of Primary Industries and Water, Hobart. <www.dpiw.tas.gov.au/inter.nsf/Attachments/JMUY-6S35Q8/$FILE/Final%20Little%20Swanport%20WMP%20June%20.pdf>.

DPIW (2006b) 'Riverine environment analysis for the Little Swanport catchment'. Water Assessment Branch, Water Resources Division. Department of Primary Industry and Water, Technical report WA 06/03, Hobart. <www.dpiw.tas.gov.au/inter.nsf/WebPages/JMUY-5C76P7?open>.

DPIW (2007a) 'Hydrophysical characterisation of the Little Swanport River catchment'. <www.dpiw.tas.gov.au/inter.nsf/Attachments/SSKA-6YT2XG/>.

DPIW (2007b) 'Little Swanport catchment: waterways monitoring report'. Water Assessment Branch, <www.dpiw.tas.gov.au/inter.nsf/Attachments/WWEB-6MJ3GP/$FILE/Little%>.

DPIW (2007c) 'The Tasmanian environmental flows framework'. Department of Primary Industries and Water, Technical Report No. WA 07/03, Hobart. <www.stors.tas.gov.au/au-7-0054-00125>.

DPIW (nd) <www.dpiw.tas.gov.au/inter.nsf/Attachments/LBUN-8GQW2E?open>.

DPIWE (2003) 'Little Swanport catchment: water resources information package'. Department of Primary Industries, Water and Environment (Water Assessment and Planning Branch), Tasmania.

DPIWE (2005a) 'Draft Little Swanport catchment water management plan. Summary of public representations'. Report prepared pursuant to S26 of the *Water Management Act 1999*. Prepared for The Resource Planning and Development Commission by Water Assessment & Planning Branch Department of Primary Industries, Water and Environment.

DPIWE (2005b) 'Environmental management goals for Tasmanian surface waters. Glamorgan-Spring Bay catchments'. Department of Primary Industries, Water and Environment, Hobart. <www.environment.tas.gov.au/file.aspx?id=1758>.

Dunlop M (2001) 'Australian water use statistics'. CSIRO Sustainable Ecosystems, Working Paper Series 01/03, Canberra.

Dyke C and Dyke S (2002) Unpublished report. Submission to national competition council re: 2002 national competition policy assessment framework for water reform.

Dyer KR (1973) *Estuaries: A physical introduction*. John Wiley and Sons, New York.

Edgar GJ, Barrett NS and Graddon DJ (1999) 'A classification of Tasmanian Estuaries and assessment of their conservation significance using ecological and physical attributes, population and land use'. Tasmanian Aquaculture and Fisheries Institute, Technical Report Series 2, Hobart.

Fasham MJR, Ducklow HW and McKelvie SM (1990) A nitrogen-based model of plankton dynamics in the oceanic mixed layer. *Journal of Marine Research* **48**, 591–639.

Garcia E, Duarte CM, Marba N, Terrados J, Kennedy H, Fortes MD and Tri NH (2003) Sediment deposition and production in SE-Asia seagrass meadows. *Estuarine, Coastal and Shelf Science* **56**, 909–919.

Global Water Partnership Technical Advisory Committee (TAC) (2000) 'Integrated water resources management'. Global Water Partnership, TAC Background Paper No. 4, Stockholm.

Gordon JDC, Boudreau PR, Mann KH, Ong JE, Silvert WL, Smith SV, Wattayakom G, Wulff F and Yanagi T (1996) Biogeochemical Modelling Guidelines, LOICZ/R&S/95-5. LOICZ, Texel, The Netherlands. <www.loicz.org>

Harris G, Nilsson C, Clementson L and Thomas D (1987) The water masses of the east coast of Tasmania – seasonal and interannual variability and the influence on phytoplankton biomass and productivity. *Australian Journal of Marine and Freshwater Research* **38**, 569–590.

Hughes L (2003) Climate change and Australia: trends, projections and impacts. *Austral Ecology* **28**, 423–443.

Hufschmidt MM, James EE, Meister AD, Bower T and Dixon JA (1983) *Environment, Natural Systems and Development: An Economic Guide*. John Hopkins University Press, Baltimore.

Hundloe T (1994) Conservation Farming in Northern Thailand. In *The Application of Economic Techniques in Environmental Impact Assessment*. (Ed. D James) pp. 143–181. Kluwer Academic Press, Dordrecht.

Hundloe T (2002) (Ed.) *Valuing Fisheries: An Economic Framework*. University of Queensland Press, St Lucia.

Hundloe, T, Morison, J, Brooks, K and Sullivan, A (2011) *Fishing for Sustainability*. Bond University Press, Robina.

International Rivers (2008) Catchment Basics. <http://www.internationalrivers.org/en/node/478>

James D (1994) *The Application of Economic Techniques in Environmental Impact Assessment.* Kluwer Academic Press, Dordrecht.

Kirkpatrick J and Bridle K (Eds) (2007) *People, Sheep and Nature Conservation: The Tasmanian Experience.* CSIRO Publishing, Melbourne.

Lawler M (2008) 'Environmental assessment of two proposed marine farming zones within Little Swanport including the summary of ADCP deployment off Point Baily'. Tasmanian Aquaculture and Fisheries Institute, Internal Report, Hobart.

Leaman D (2004) Management of the Little Swanport River: with particular reference to a dam at Stonehenge. Expert Opinion. Leaman Geophysics, Bellerive, Tasmania.

Leaman D (2005) *Water: Facts, Issues and Problems.* Leaman Geophysics, Hobart.

Leaman D (2007) *Water: Facts, Issues, Problems and Solutions.* 2nd edn. Leaman Geophysics, Hobart.

Lewis D (2006) Modern and recent seafloor environments (sedimentary, foraminiferal and Ostracode) of the Pitt Water Estuary, south-east Tasmania. PhD thesis. The University of Tasmania, Australia.

Little Swanport Catchment Committee (2006) 'Little Swanport catchment water management plan'. June 2006. Report compiled by Little Swanport Catchment Committee. Water Policy and Planning Branch, Department of Primary Industries and Water, Hobart.

Lourandos H (1970) Coast and hinterland. The archaeological sites of eastern Tasmania. MA thesis. Australian National University, Australia.

Madden CJ and Kemp WM (1996) Ecosystem model of an estuarine submersed plant community: calibration and simulation of eutrophication responses. *Estuaries* **19**, 457–474.

Mann KH and Lazier JRN (2006) *Dynamics of Marine Ecosystems: Biological-Physical Interactions in the Oceans.* Blackwell Publishing, Oxford.

Mann R (1979) Some biochemical and physiological aspects of growth and gametogenesis in Crassostrea-Gigas and Ostrea-Edulis grown at sustained elevated-temperatures. *Journal of the Marine Biological Association of the United Kingdom* **59**, 95–110.

Markwick G (2002) 'Water requirements for sheep and cattle'. Agfact A0.5.4. DPI NSW Agriculture. <www.dpi.nsw.gov.au/__data/assets/pdf_file/0009/96273/water-requirements-for-sheep-and-cattle.pdf>.

Martin R and Procter C (2000) 'Tasmanian Marine Farm Marine Pest Monitoring Project'. Centre for Research on Introduced Marine Pests, CSIRO Marine Research, Final report to NHT/C&CS, Hobart.

McIntosh PD, Kiernan K and Price DM (2004) An aeolian sediment pulse at C.28Kyr BP in southern Tasmania. *Journal of the Royal Society of New Zealand* **34**, 369–379.

McIntosh PD, Laffin MD and Hewitt AE (2005) The role of fire and nutrient loss in the genesis of forest soils of Tasmania and southern New Zealand. *Forest Ecology and Management* **220**, 185–215.

Micevski T, Franks SW and Kuczera G (2006) Multidecadal variability in coastal eastern Australian flood data. *Journal of Hydrology*, **327**, 219–225.

Miller GT (2007) *Living in the Environment.* 14th edn. Brooks/Cole Publishing, Belmont, California.

Mount M, Lucieer V, Lawler M and Jordan A (2005) 'Mapping of estuarine and marine habitats in the southern NRM region'. Tasmanian Aquaculture and Fisheries Institute, Final report to NHT, Hobart. <www.utas.edu.au/tafi/seamap/>

Mulvaney J and Kamminga J (1999) *Prehistory of Australia.* Allen and Unwin, Sydney.

Murphy R, Crawford C and Barmuta L (2003) 'Estuarine health in Tasmania, status and indicators: water quality'. Tasmanian Aquaculture and Fisheries Institute, Technical Report Series 16, Hobart.

National Water Commission (2008) 'Australian Water Resources 2005'. Australian Government. <www.water.gov.au/publications/index.aspx?Menu=Level1_9>

National Heritage Trust (2004) National Land and Water Resources Audit. <www.nlwra.gov.au>.

Nyman L (1990) *The East Coasters: The Early Pioneering History of the East Coast of Tasmania.* Regal Publications, Launceston.

Parliamentary Report (1883) Fisheries of Tasmania: Report of Royal Commission; 1882, No. 92 Legislative Council (132 House of Assembly), Hobart.

Pinto R (2001) 'Environmental Water Requirements for the Little Swanport river'. Department of Primary Industries, Water and Environment, Report series WRA 01/08, Hobart.

Pittock B (2003) *Climate change: An Australian guide to the science and potential inpacts.* Australian Greenhouse Office, Canberra.

Potts J (2004) 'Quantification of sediment-water nutrient and oxygen flux rates in Little Swanport Inlet, Tasmania'. Department of Environment and Conservation(NSW), Internal Report, Sydney.

Ren JS, Marsden ID, Ross AH and Schiel DR (2003) Seasonal variation in the reproductive activity and biochemical composition of the Pacific oyster (*Crassostrea gigas*) from the Marlborough Sounds, New Zealand. *New Zealand Journal of Marine and Freshwater Research* **37**, 171–182.

Resource Planning and Development Committee (2003) 'State of the environment: case studies. Little Swanport River Catchment'. <http://soer.justice.tas.gov.au/2003/casestudy/8/index.php>. Accessed 30 June 2008.

Resource Planning and Development Commission (2006) 'Report on the Draft Little Swanport Catchment Water Management Plan 2004'. May 2006. Resource Planning and Development Commission, Hobart.

Robson LL (1983) *A history of Tasmania. Volume 1. Van Diemen's Land from the earliest times to 1855.* Oxford University Press, Melbourne.

Shaw E (2000) *My Memoirs Laced With East Coast Tales of Van Diemens Land (Tasmania).* Edward C. Shaw, Tasmania.

Sinclair Knight Merz (2004) 'Little Swanport River catchment: water balance model and scenario assessment'. Sinclair Knight Merz, Armadale, Victoria.

Switzer AD and Jones BG (2008) Large scale washover sedimentation in a freshwater lagoon from the southeast Australian coast: sea-level change, tsunami or exceptional large storm. *The Holocene* **18**, 787–803.

Taylor A (1892) Notes on the shell mounds at Seaford, Little Swanport. *Papers and Proceedings of the Royal Society of Tasmania.* Royal Society of Tasmania, Hobart, pp. 89–94.

Temby N and Crawford C (2008) 'Coastal and estuarine resource condition assessment: A baseline survey in the Southern NRM Region, Tasmania'. Tasmanian Aquaculture and Fisheries Institute, Final report to NHT, Hobart.

United Nations (2003) 'SEEA Integrated Environmental and Economic Accounting 2003'. Studies in Methods Handbook of National Accounting. Series F, No. 61, Rev. 1. New York.

United Nations (2006) *Draft Handbook on Environmental and Economic Accounting for Water Resources.* United Nations Statistical Division, New York.

Vardon M, Lenzen M, Peevora S and Creasera M (2007) Water accounting in Australia. *Ecological Economics* **61(4)**, 650–659.

von Platen J (2004) Interpreting fire regimes in dry eucalyptus forests using tree ring chronologies in the eastern tiers of Tasmania. *School of Geography and Environmental Studies 2004 Conference Extracts*, October 2004. University of Tasmania, Hobart, p. 18. <http://ecite.utas.edu.au/30319>

von Platen J (2008) A history and interpretation of fire frequency in dry eucalypt forests and woodlands of Eastern Tasmania. PhD thesis. The University of Tasmania, Australia.

Webster IT, Parslow JS and Smith SV (2000) Implications of spatial and temporal variation for biogeochemical budgets of estuaries. *Estuaries* **23**, 341–350.

Wilson RS, Cohen BF and Poore GCB (1993) 'The role of suspension-feeding and deposit-feeding benthic macroinvertebrates in nutrient cycling in Port Phillip Bay'. Museum of Victoria, Report No. 10, Melbourne.

Index

www.ingramcontent.com/pod-product-compliance
Ingram Content Group UK Ltd.
Pitfield, Milton Keynes, MK11 3LW, UK
UKHW062009041026
15591UKWH00017B/1243